Applied Neural Networks and Fuzzy Logic in Power Electronics, Motor Drives, Renewable Energy Systems and Smart Grids

Applied Neural Networks and Fuzzy Logic in Power Electronics, Motor Drives, Renewable Energy Systems and Smart Grids

Editors

Marcelo Godoy Simões
Helmo Kelis Morales Paredes

MDPI • Basel • Beijing • Wuhan • Barcelona • Belgrade • Manchester • Tokyo • Cluj • Tianjin

Editors
Marcelo Godoy Simões
Colorado School of Mines
USA

Helmo Kelis Morales Paredes
São Paulo State University—UNESP
Brazil

Editorial Office
MDPI
St. Alban-Anlage 66
4052 Basel, Switzerland

This is a reprint of articles from the Special Issue published online in the open access journal *Energies* (ISSN 1996-1073) (available at: https://www.mdpi.com/journal/energies/special_issues/Fuzzy_Logic_Smart_Grids).

For citation purposes, cite each article independently as indicated on the article page online and as indicated below:

LastName, A.A.; LastName, B.B.; LastName, C.C. Article Title. *Journal Name* **Year**, *Article Number*, Page Range.

ISBN 978-3-03943-334-6 (Hbk)
ISBN 978-3-03943-335-3 (PDF)

Contents

About the Editors

Marcelo Godoy Simões received the B.Sc. and M.Sc. degrees from the University of São Paulo, Sorocaba, SP, Brazil, and his Ph.D. degree from The University of Tennessee, Knoxville, TN, USA, and the D.Sc. degree from the University of São Paulo, in 1985, 1990, 1995, and 1998, respectively. He was a US Fulbright Fellow during the period AY 2014–2015, working with the Institute of Energy Technology, Aalborg University, Aalborg, Denmark. He is currently serving as a Faculty Senator with the Colorado School of Mines, Golden, CO, USA, and as a Vice-President for the American Association of University Professors Chapter with the Colorado School of Mines. He is a pioneer to apply neural networks and fuzzy logic in power electronics, motor drives, and renewable energy systems. His fuzzy logic-based modeling and control for wind turbine optimization is used as a basis for advanced wind turbine control and it has been cited worldwide. His leadership in modeling fuel cells is internationally and highly influential in providing a basis for further developments in fuel cell automation control in many engineering applications. He made a substantial and lasting contribution of artificial intelligence technology in many applications, power electronics and motor drives, fuzzy control of a wind generation system, such as fuzzy logic-based waveform estimation for power quality, neural network-based estimation for vector-controlled motor drives, and the integration of alternative energy systems to the electric grid through artificial intelligence (AI) modeling-based power electronics control.

Dr. Simões was a recipient of the "2018 IET Renewable Power Generation Premium Award." He is an IEEE Fellow, Class of 2016, with the citation: "for applications of artificial intelligence in control of power electronics systems." He is currently serving as the Chair for the IEEE Industrial Electronics Society Denver Chapter.

Helmo Kelis Morales Paredes received the B.S. degree from the San Agustin National University, Arequipa, Peru, in 2002, and his M.Sc. and Ph.D. degrees from the University of Campinas, Campinas, Brazil, in 2006 and 2011, respectively, all in electrical engineering. In 2009, he joined the Power Electronics Group, University of Padova, Padova, Italy, as a Visiting Student. In 2014, he joined the PEMC Group, University of Nottingham, U.K., as a Visiting Scholar. In 2018, he joined the Colorado School of Mines, Golden, CO, USA, as a Visiting Scholar. Since 2011, he has been with São Paulo State University—UNESP, as an Associated Professor and he is the Leader of the Group of Automation and Integrating Systems. His current research interests include power theories under non-sinusoidal condition, harmonics propagation, power quality, accountability, revenue metering, and power electronics applied for distributed generation and renewable energy systems.

Dr. Morales-Paredes was a recipient of a Prize Paper Award from the IEEE Transaction on Power Electronics in 2011 and currently he is an Associate Editor for the IEEE Latin America Transactions. He is a member of the Brazilian Power Electronic Society, Brazilian Automation society and IEEE.

Preface to "Applied Neural Networks and Fuzzy Logic in Power Electronics, Motor Drives, Renewable Energy Systems and Smart Grids"

This Special Issue addresses the latest results on various aspects of artificial intelligent techniques, such as expert systems, fuzzy logic, and artificial neural networks in important areas of advanced high-tech electronics, such as applications in power electronics, motor drives, renewable energy systems and smart grids. There is a multidisciplinary relevant to all of those. Through this framework, based on fuzzy logic, an integrated hierarchical data envelopment analysis for the optimal geographical location of solar plants is proposed, a distributed multi-source information is obtained for equipment condition assessment in the power grid, and control methods to regulate the demand of thermostatically controlled loads and wind turbine systems are also proposed to provide frequency support for a smart grid. In addition, using deep learning algorithm transient voltage stability analysis to optimize the reactive power compensation and detect non-technical losses is proposed and evaluated. Finally, hourly electricity load demand, multiple electricity consumption forecasting and non-intrusive load disaggregation methodology are also presented in this Special Issue.

Marcelo Godoy Simões, Helmo Kelis Morales Paredes
Editors

Article

Frequency Support of Smart Grid Using Fuzzy Logic-Based Controller for Wind Energy Systems

Marcelo Godoy Simões [1,*] and Abdullah Bubshait [2]

1   Electrical Engineering Department, Colorado School of Mines, Golden, CO 80401, USA
2   Electrical Engineering Department, King Faisal University, Alahsa 31982, Saudi Arabia; asbubshait@kfu.edu.sa
*   Correspondence: msimoes@mines.edu; Tel.: +1-303-384-2350

Received: 22 February 2019; Accepted: 19 April 2019; Published: 24 April 2019

Abstract: This paper proposes a fuzzy logic-based controller for a wind turbine system to provide frequency support for a smart grid. The designed controller is aimed to provide an appropriate dynamic droop rate depending on the local measurements of each wind turbine of a wind farm such as the maximum power available and the amount of power reserve. The designed fuzzy controller depends on the rate of change of frequency (ROCOF) at the point of common coupling (PCC). The main advantage of the proposed fuzzy controller is to provide frequency support by the wind turbine system connected to a smart grid. The dynamic rate of the controller is defined by the fuzzy sets considering the change in the grid's frequency and the available reserve power. First, the response of static droop curves is investigated for different scenarios of wind turbines connected to a smart grid. Then, the proposed fuzzy logic-based droop controller is integrated into the system, and its performance and response are evaluated, and the results are compared with static-droop based controller. The proposed controller is tested using Matlab\Simulink.

Keywords: droop curve; frequency regulation; fuzzy logic; the rate of change of frequency; reserve power; smart grid

---

1. Introduction

Conventional synchronous generators store kinetic energy in their rotor shaft and provide inertia to the system. The frequency of the power system is directly coupled to the rotational speed of the generator. Therefore, more rotational inertia may lead to a stable power system. Recently, the installation of renewable resources in the power system has increased. These resources are integrated into the power system using power converters, but they do not provide rotating inertia to the power system. High penetration of distributed energy resources (DER) might create instability if not properly controlled. Therefore, ancillary functions had to be added and must be achieved by the controllers of the DER. Renewable resources can provide voltage regulation and frequency support if controlled properly.

For instance, the wind energy system (WES) can be adjusted to provide frequency support by controlling the injected active power as a function of frequency. To do so, WES must maintain a certain reserve of active power, and then the reserved power can be utilized during a frequency drop.

The active power control (APC) during a disturbing condition is a challenge when it is applied to a wind farm. The APC requires a fast dynamic response range in a few seconds. Active power control involves three main sub-control objectives in a power system: the inertial control, the primary frequency control (PFC), and automatic generation control (AGC) [1,2]. Each is involved in the control of active power for a certain time in sequence order due to their timely response. One main challenge is that wind availability is uncertain, causing failure in the control loop due to wind speed variation.

Participating in frequency control can be implemented by adjusting the conventional control techniques of WES systems [1–14]. Different control methods, and approaches have been implemented to adjust the frequency of a power system during frequency fluctuation [15,16]. These control strategies can be classified based on the capability and duration of participation [15–19]. The first approach is to use the stored kinetic energy to be implemented as inertial control. The stored energy can be released for a few seconds depending on the inertia of the turbine. The other control is to de-load the wind turbine below the maximum power point, providing a long-term power supply following the inertial response to maintain a new steady state value of frequency. This is known as a primary frequency response (PFC). There have been several studies on implementing the PFC in power systems with existing wind turbines. Some studies have focused on the wind farm level, whereas others have focused on the power system level [20–23]. Also, some studies have analyzed the equivalent damage loads for de-loaded wind turbines [24]. Other researchers have focused on the reserve methods and the response of the wind turbines to PFC [25–28].

In reference [29], the active power control strategy was performed for a WES using inertial, PFC and AGC. Different droop control approaches were proposed to support the frequency of the grid. Variable droop controller was proposed to enhance the response of WES based on doubly fed induction generator DFIG in reference [30]. The concept of droop can be implemented in the local controller of the WES, at the wind farm level, or to the coordinated distributed generator in the smart grid [31–40].

In this article, a fuzzy logic-based controller is developed for the WES to provide frequency support for a smart grid. The controller is designed to identify the participation factor of each wind turbine based on the reserved power and the ROCOF at the point of common coupling (PCC).

A power system is developed to test the response of the wind farm to the frequency drop.

2. Stability of the Power Grid

A power system is a complicated structure involving various elements with different dynamics. In the ideal world, loads of the power system are provided with consistent frequency and voltage. At normal conditions, all synchronous generators are synchronized to avoid up-normal fluctuation in the voltage and current, which may lead to disconnecting areas from the grid. The frequency deviation is a result of an imbalance between the load and power generated. The frequency of the gird is related to the rotor speed of the synchronous generators of the power system. The change of frequency of a network can be given as [41]:

$$\Delta f = \frac{1}{\left(2H_{sys}\right)S + D_{sys}}\left(\sum \Delta P_G - \sum \Delta P_L\right) \tag{1}$$

where H_{sys} and D_{sys} are the equivalent inertia and damping constants of all machines of the system. The change in generated power and load power are represented by ΔP_G and ΔP_L respectively. To ensure stability, the synchronous generator should be equipped with a droop curve in the speed controller of the turbine's torque-speed characteristic. The droop characteristic is typically set to 5% in the United States [42].

3. Frequency Support by a Wind Energy System

WES can be controlled to maintain certain reserve power and then, it can be used to provide APC by modifying the control loop to follow the required power reference. To do so, a droop control concept is implemented where the active power is related to the frequency of the grid.

The required active power to stabilize the grid frequency can be achieved by implementing the droop curve. The response of WES during frequency drop depends substantially on the de-loading method used to maintain the reserve power (i.e., rotor speed or pitch angle). Also, the response of the WES can be impacted by the initial operating points of WES just slightly before the occurrence of frequency deviation. Thus, the droop curve has to be selected cautiously to guarantee stability and reliable response of WES. For instance, very fast droop for a de-loaded WES that operates away from its maximum point may lead to controller instability [43,44].

The droop curve that implemented in the synchronous generator can be adapted to be used for WES. The variation in the active power of a WES can be expressed as [35]:

$$P_{PFC} = \frac{P_{WT,ava}}{R} \frac{\left(f_{nom.} - f_{grid}\right)}{f_{nom.}} \tag{2}$$

where $P_{WT,ava}$ is the maximum power available that can be generated by the WES. f_{grid} is the measured grid frequency; and $f_{nom.}$ the nominal grid frequency are represented respectively. Here, R is defined as the slope of the droop curve represented in percentage. The slope determines the rate of power change in WES. Small droop rate means a fast change in active power. Figure 1 demonstrates different droop curves as function of an active power change in percentage.

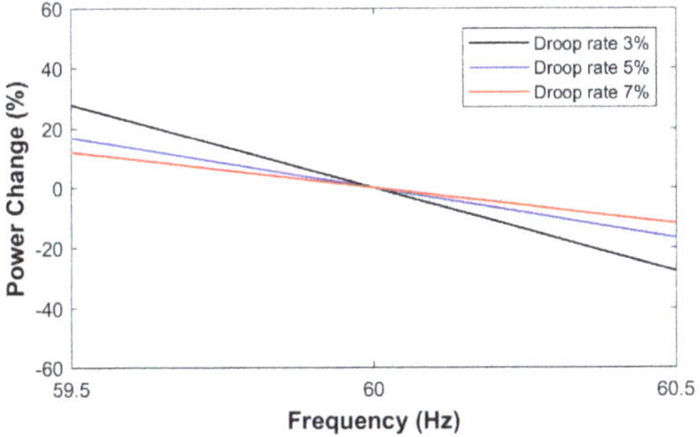

Figure 1. The change of power for different droop rates.

The rate of the droop controller must be selected to ensure a reliable and smooth response against the deviation of the grid's frequency. Designing the droop curve for frequency regulation depends on the initial state and the power availability of the WES. Because of the variation of wind speed, the active power produced by WES is variable. As a result, the amount of the power reserve for WES is also dynamic. Also, the ROCOF at the PCC varies depending on how much power is lost from the network. Therefore, for WES a dynamic controller is the best fit to provide adjustment to the grid's frequency.

4. Fuzzy Logic-Based Controller

A dynamic droop controller based on fuzzy logic is designed to support the grid's frequency of WES. The fuzzy logic controller is a set of rules are defined to perform a certain function [45]. The concept of fuzzy logic has been used in control applications of the electrical grid and power electronics [46–50].

Fuzzy logic is implemented to determine the slope of the droop curve depending on the available reserved power of WES and the ROCOF. For instance, when the change of frequency is high, and the power reserve is not large, the WES must support the frequency by providing active power with a slow rate. In the other hand, when the power reserve is high, WES should support the frequency by rapidly providing the required active power.

The equation that relates the grid's frequency and active power are:

$$P_{PFC} = \frac{P_{WT,ava}}{R_t} \frac{\left(f_{nom.} - f_{grid}\right)}{f_{nom.}} \tag{3}$$

This fuzzy logic-based controller aims to give an appropriate dynamic rate (R_t). The output of the fuzzy logic controller depends on two inputs. The inputs are the ROCOF, and the power reserved by the

WES (ΔP_{WT}) at the time of frequency drop. The reserved power of WES varies according to the wind velocity and the de-loading method used by the central controller of a wind plant. Also, the wake-effect plays a vital role on the power availability for each individual WES. Therefore, each WES has a different power reserve to others.

Figure 2 demonstrates the implementation of fuzzy-logic with the dynamic droop controller. Normally, the wind speed, V_W, can be measured to estimate the maximum power of the turbine, $P_{WT,\,max}$. The maximum power is decreased by ΔP_{WT} to maintain a certain reserve. Every individual WES is de-loaded to maintain a certain reserve of power (ΔP_{WT}) to be utilized to provide APC during a drop in the grid's frequency. When grid's frequency drops below its set point, the controller injects a certain amount of active power into the grid. Then, this power, PFC, is added to the loop to produced power to form a reference of the required total wind power. The reference point, P_{WT}, of the total power is then achieved either by using a pitch angle or rotor speed controller.

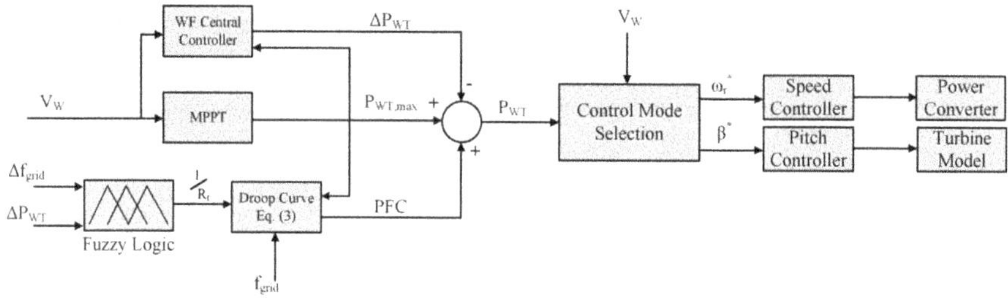

Figure 2. The implementation of fuzzy logic within the active power controller.

4.1. Fuzzification

The crisp values of the two inputs are mapped into fuzzy sets using the triangle membership function. Both inputs are defined using five fuzzy sets as shown in Figures 3 and 4. The sets of the inputs are defined as: very large (VL), large (L), medium (m), small (S), and very small (VS).

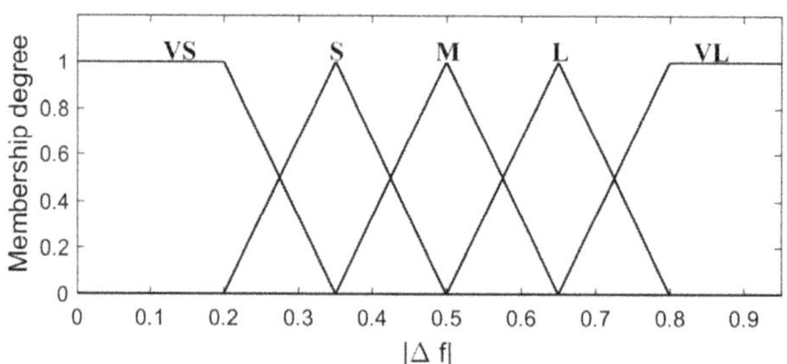

Figure 3. Membership function (changing in grid's frequency).

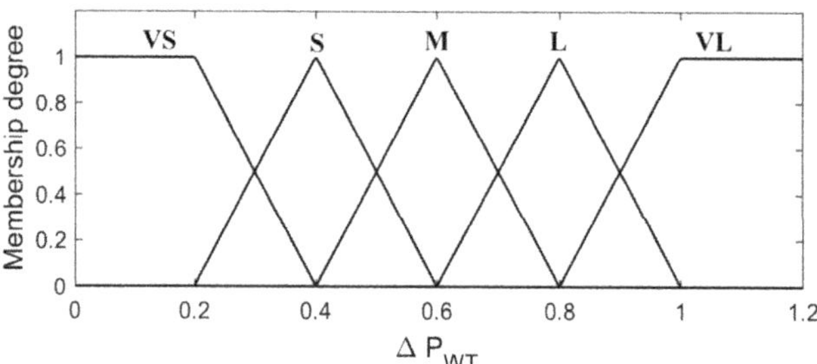

Figure 4. Membership function (power reserve).

The absolute value of the rate of change in the grid's frequency is represented by the first membership (Figure 3), whereas the available reserved power is represented by the second membership function (Figure 4). In this paper, frequency drop less than 0.2 Hz is defined by the fuzzy sets as very small (VS) and above 0.8 Hz is defined as very large (VL).

In this article, wind energy systems rated at 2 MW are considered. In these wind energy systems, the amount of reserve can range from 100 kW to 1 MW (i.e., 5% to 50% of the maximum power). For simplification, the power reserve is measured and scaled down before it enters the membership function. The dynamic rate is given by the output membership function shown in Figure 5. The output is determined by the following fuzzy sets: very fast (VF), fast (F), medium (M), slow (SL) and very slow (VSL). This output is then used to determine the rate of change in injected active power given in (3). The rules of the fuzzy logic used to define the output are discussed in the following section.

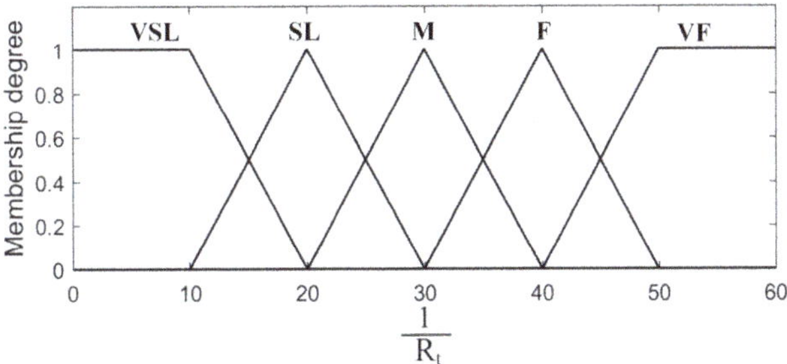

Figure 5. Membership function (output).

4.2. Fuzzy Inference Rules

The output of the fuzzy logic is determined by defining 25 rules, indicated in Table 1. The fuzzy inference rules are based on the deviation in frequency (Δf) and the amount of reserve (ΔP_{WT}). To evaluate the inference rules of the sets, the minimum conjunction operator is used.

Table 1. Fuzzy inference rules.

$\Delta P_{WT}\backslash\Delta f$	VS	S	M	L	VL
VS	VSL	VSL	SL	M	M
S	VSL	SL	M	F	M
M	SL	M	M	F	F
L	M	M	F	F	VF
VL	M	M	VF	VF	VF

Fast rate is desired if the frequency drop and the amount of reserve of WES are large. If the amount of reserved power of the WES is not large and the ROCOF is small, the low rate is preferred. The fuzzy inference rules are defined to avoid an unnecessary fast rate for very small reserves. The fuzzy inference rules are explained below:

- IF ΔP_{WT} is VS AND Δf is VS THEN R_t is VSL
- IF ΔP_{WT} is VS AND Δf is S THEN R_t is VSL
- IF ΔP_{WT} is VS AND Δf is M THEN R_t is SL
- IF ΔP_{WT} is VS AND Δf is L THEN R_t is M
- IF ΔP_{WT} is VS AND Δf is VL THEN R_t is M
- IF ΔP_{WT} is S AND Δf is VS THEN R_t is VSL
- IF ΔP_{WT} is S AND Δf is S THEN R_t is SL
- IF ΔP_{WT} is S AND Δf is M THEN R_t is M

- IF ΔP_{WT} is S AND Δf is L THEN R_t is F
- IF ΔP_{WT} is S AND Δf is VL THEN R_t is F
- IF ΔP_{WT} is M AND Δf is VS THEN R_t is SL
- IF ΔP_{WT} is M AND Δf is S THEN R_t is M
- IF ΔP_{WT} is M AND Δf is M THEN R_t is M
- IF ΔP_{WT} is M AND Δf is L THEN R_t is F
- IF ΔP_{WT} is M AND Δf is VL THEN R_t is F
- IF ΔP_{WT} is L AND Δf is VS THEN R_t is M
- IF ΔP_{WT} is L AND Δf is S THEN R_t is M
- IF ΔP_{WT} is L AND Δf is M THEN R_t is F
- IF ΔP_{WT} is L AND Δf is L THEN R_t is F
- IF ΔP_{WT} is L AND Δf is VL THEN R_t is VF
- IF ΔP_{WT} is VL AND Δf is VS THEN R_t is M
- IF ΔP_{WT} is VL AND Δf is S THEN R_t is M
- IF ΔP_{WT} is VL AND Δf is M THEN R_t is VF
- IF ΔP_{WT} is VL AND Δf is L THEN R_t is VF
- IF ΔP_{WT} is VL AND Δf is VL THEN R_t is VF

4.3. Defuzzification

The fuzzy inference rules give linguistic variables that need to be transformed into crisp values. Therefore, the defuzzification process is implemented to quantify the output of the fuzzy logic. To do so, there are different methods of defuzzification. In this article, the weighted average method is implemented. The defuzzified slope rate (i.e., $1/R_t$) is defined as:

$$output = \frac{\sum_{i=1}^{m} Z_i * \mu(Z_i)}{\sum_{i}^{m} \mu(Z_i)} \tag{4}$$

where Z_i represents the center of the defined function, and $\mu(Z_i)$ is the value of the membership that corresponds to Z_i. Then, the output ($1/R_t$) of the fuzzy-logic controller is used to gives the rate of the droop in Equation (3).

5. Simulation Results

To study the performance of the proposed controller, the small power system is considered as shown in Figure 6. The model of the wind turbine considered in this paper is defined as type-4, where the WT is connected to the grid through a full rated back-to-back converter. The machine used in this system is based on a permanent magnet synchronous generator (PMSG). The machine side converter is controlled to extract the desired power from the wind. The active and reactive power is controlled using the grid-side inverter. The detailed model of the implemented WT system and its designed controller are presented in reference [51].

In the first study, one single WES was used to test the performance of the fuzzy logic-based controller. Different static droop curves were tested and compared with the proposed approach. The WES was set to regulate the frequency of the grid by providing the required power. The loads in the power system are supplied by the synchronous generator and the WES. The conventional synchronous generator of the system is rated at 15 MW. Its contribution to the total power of the grid is 90%. The rating of the WES used in the simulation is 2 MW. It represents about 10% of the total power delivered to the loads.

After that, the simulation was repeated for the same power system model using a wind farm, a conventional synchronous generator, and load. In this study, the wind farm consists of four single WESs. Here, the synchronous generator provides 80% of the load power, while the wind farm share is

20%. The wind systems are de-loaded to maintain a certain power reserve to be utilized for frequency regulation. Due to the wake effect, the produced wind power and the reserves are not similar for all wind systems. Therefore, some wind systems may have a much higher power reserve than others. Different scenarios were performed to test the response of the wind farm using several droop rates.

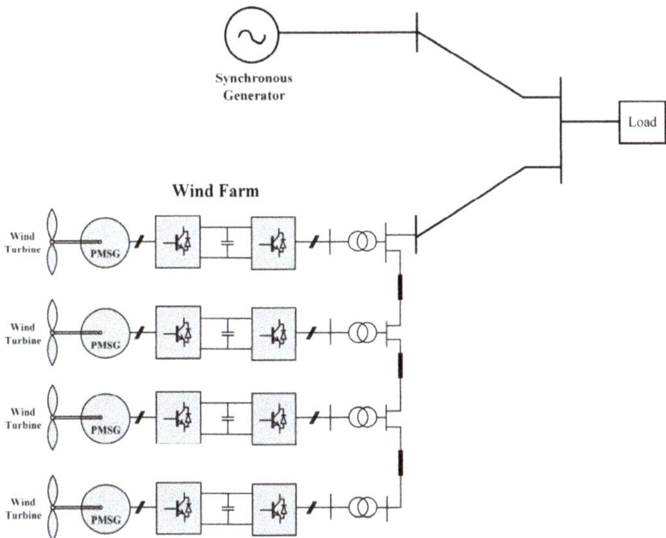

Figure 6. Power system model with a wind farm of 4 wind turbine systems.

5.1. Frequency Support by Single WES

To study the response of wind turbines to frequency deviation, several simulation-based studies were performed. The goal was to compare the responses of different droop curves for different ROCOF. The sensitivity to different droop gains was studied and compared with different control approaches.

5.1.1. The Response of WES to Large ROCOF

For this case study, the WES was producing its maximum power (2.0 MW). The power reserve of the WES was set to 1.0 MW (i.e., a 50% rate). First, the frequency regulation was achieved using the proposed fuzzy-logic controller. Then, the simulation was repeated using two different static droop rates ($R = 2\%$, 7%). Figure 7 demonstrates the system frequency of the baseline simulation (without WES participation in frequency regulation). Figure 8 shows the frequency of the grid for the three simulation studies (fuzzy logic, $R = 2\%$, 7%). The proposed fuzzy logic-based controller provides an acceptable frequency support if compared to both static droop curves.

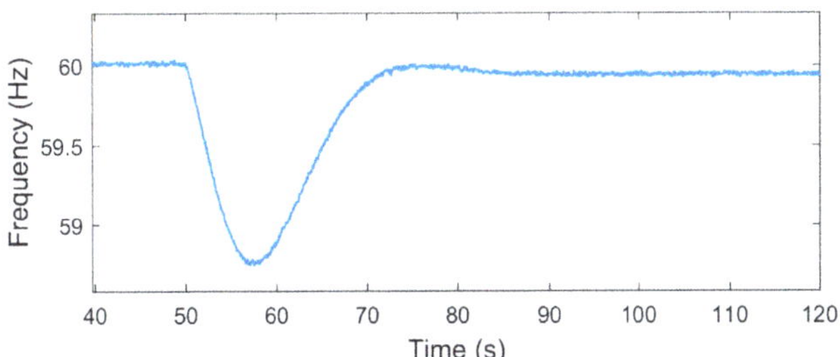

Figure 7. Grid frequency of the event without wind turbine participation.

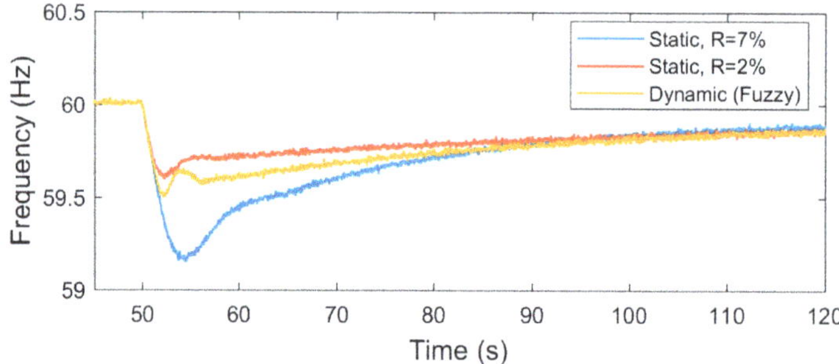

Figure 8. Grid frequency (large ROCOF).

The rate of dynamic droop starts at 10% and then decreases to about 3% at the beginning of the incident. After a few seconds, it reaches 2% (very fast), when the frequency drops to its minimum value as shown in Figure 9. For 7% static droop, the WES has a very slow response. On the other hand, the response of WES with 2% static droop is very fast.

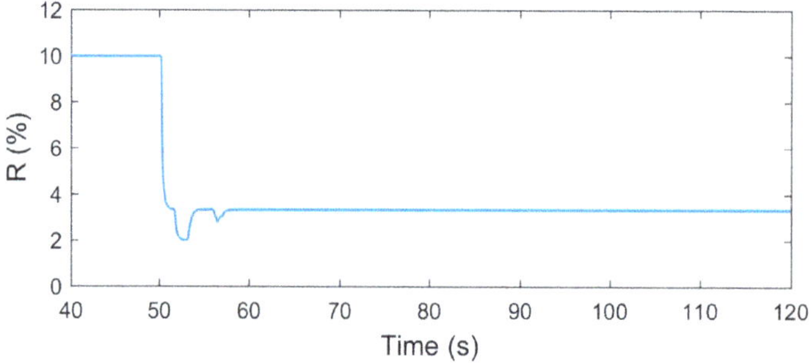

Figure 9. Dynamic droop (large ROCOF).

As defined in the fuzzy logic, the dynamic droop controller should react very quickly when the power reserve and ROCOF are large. Consequently, the response of the WES with dynamic droop is quite similar to the one with the static droop curve of 2%. The active power of the WES and the rotation speed of the turbine's shaft during the response to frequency drop are shown in Figures 10 and 11.

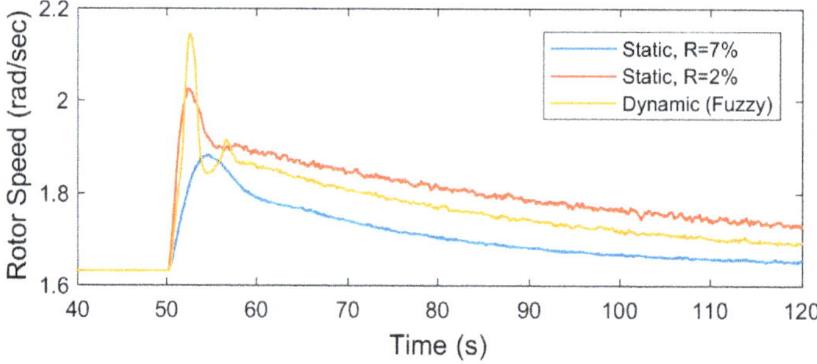

Figure 10. Rotor speed (large ROCOF).

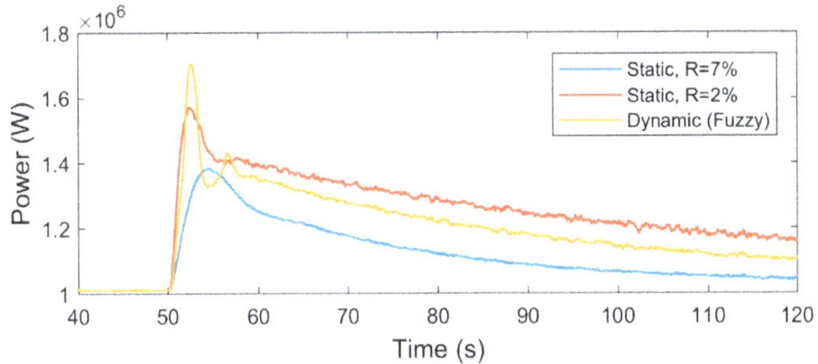

Figure 11. Wind turbine power (large ROCOF).

5.1.2. Response of WES to Small ROCOF

In this scenario, the WES was maintaining a reserve of 150 kW. This reserve was achieved by the controller of the rotor speed (using machine side converter). For comparison, the frequency support was provided using the proposed approach (fuzzy-logic) and two constant rate droops (i.e., $R = 2\%$ and 7%).

The measured frequency of the power system for the three droops (dynamic droop, $R = 2\%$ and 7%) is shown in Figure 12. The plot shows the proposed droop controller supports the frequency. The rate of the proposed droop controller is shown in Figure 13. The rate starts at the minimum value, which is 10%, and changes to about 5% in the beginning of the frequency drop. After that, it oscillates and returns to 10% when a new steady state value is achieved.

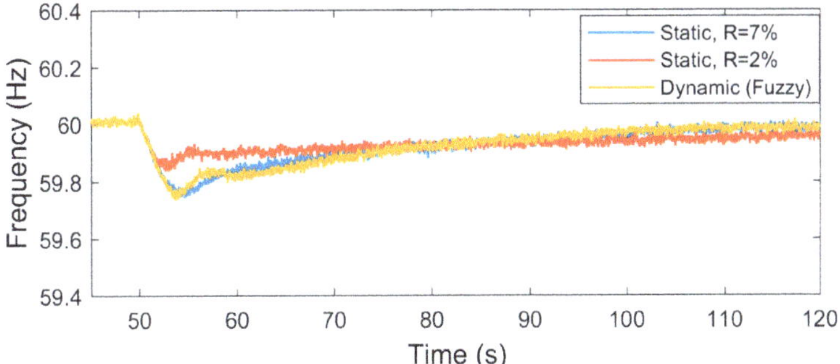

Figure 12. Grid frequency (small ROCOF).

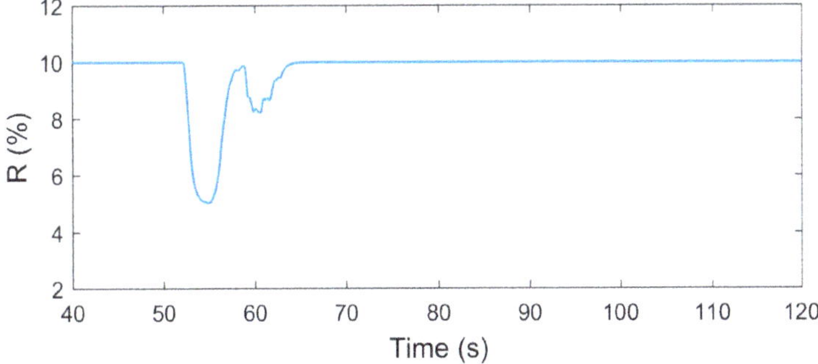

Figure 13. Dynamic droop (small ROCOF).

In this simulation, the response of the WES using constant droop of 7% is reasonable. In contrast, the response with a constant curve of 2% is quick. As can be noticed from Figure 14, the rotor speed of the WES oscillates aggressively. This high oscillation may lead to instability in the controller of the machine side converter. Also, some stresses and mechanical loading can be observed on the wind turbine. Because of the limited reserve and insignificant drop in the frequency, the droop rate of the controller has to be slow. Thus, the grid frequency can be achieved by the proposed droop controller, which in this case study is very close to the response of seven percent droop curve. The active power produced by the WES during the frequency regulation for all three studies is demonstrated in Figure 15.

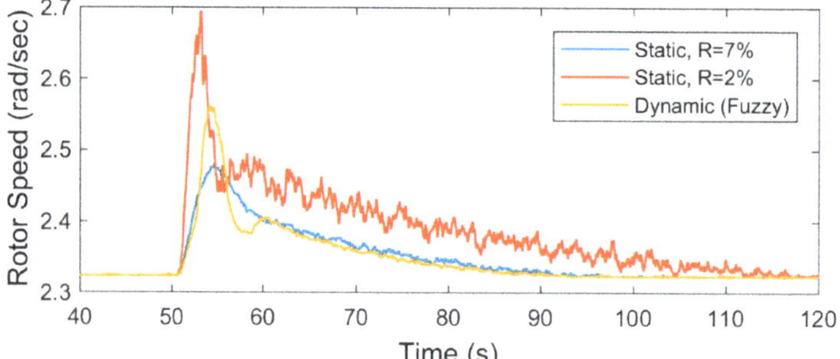

Figure 14. Rotor speed (small ROCOF).

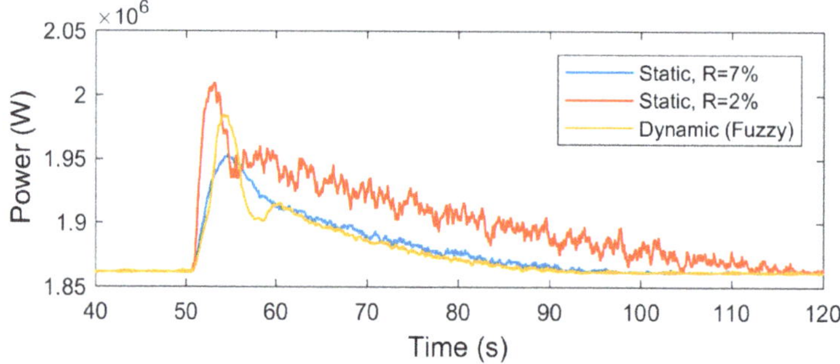

Figure 15. Wind turbine power (small ROCOF).

5.2. Frequency Support by Wind Farm

In this study, an electrical system shown in Figure 6 is simulated. The electrical grid consists of one conventional synchronous generator with its governor system, wind power plant, and a large load. For the simulation time constraint, four single WESs are considered to represent the wind farm. The power supplied to the load is distributed between the wind farm (20%) and the conventional generator (80%).

The response of every WES was observed using the proposed controller. A reserve of 2000 kW was maintained by the wind farm using a method proposed in reference [52]. The fuzzy logic-based controller was implemented to provide frequency support. Every individual WES was assigned to maintain a certain reserve that is different from others. The total power reserve was divided among all WESs as $\Delta P_{WT1} = 250$ kW, $\Delta P_{WT2} = 350$ kW, $\Delta P_{WT3} = 600$ kW, $\Delta P_{WT4} = 800$ kW.

The frequency is measured at the point as shown in Figure 16. The proposed dynamic controller provides support to the grid's frequency. The frequency goes to the minimum point (nadir) within 3 s and then it starts to return to new steady-state value. The rates of the dynamic controller for all WES are shown in Figure 17. The rates of wind turbines 1 and 2 are slower than the rates of wind turbines 3 and 4; because of the small amount of reserve available in 1 and 2. The power produced by the synchronous generator and the wind farm is shown in Figures 18 and 19. The power produced by each WES is demonstrated in Figure 18. The power of wind turbines 3 and 4 changes rapidly compared to the power of wind turbines 1 and 2 as shown in Figure 19. Also, the rotational speed of each turbine is shown in Figure 20. In this study, the controller of the rotor speed and the pitch angle actuator are activated, and the WES is tracking the reference signal given by the fuzzy logic controller as demonstrated in Figure 21.

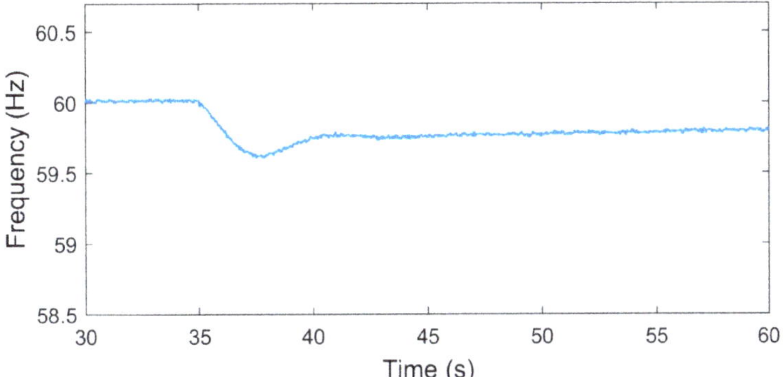

Figure 16. Grid frequency (wind farm).

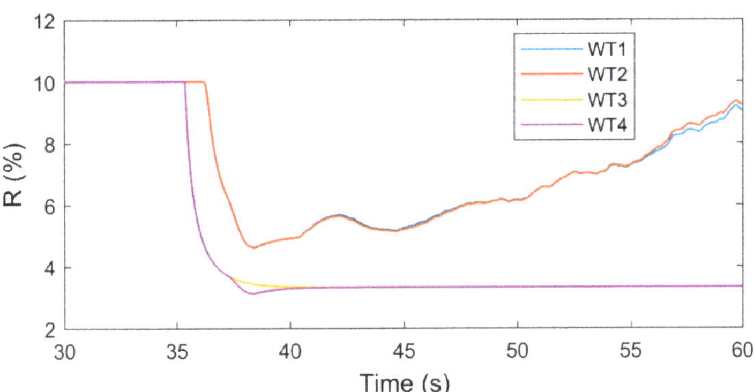

Figure 17. Dynamic droop (wind farm).

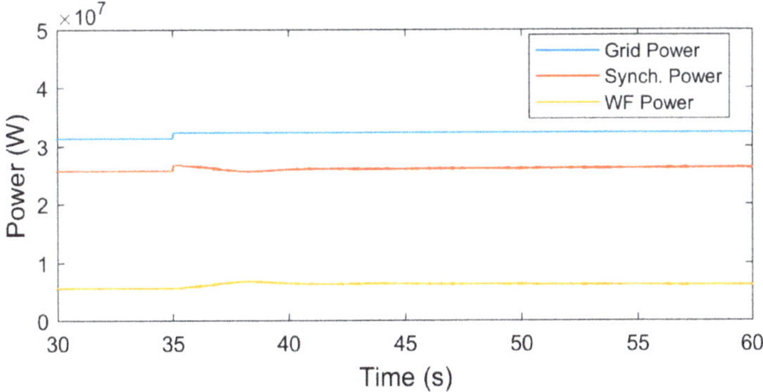

Figure 18. Power: grid, synchronous generator, and wind farm.

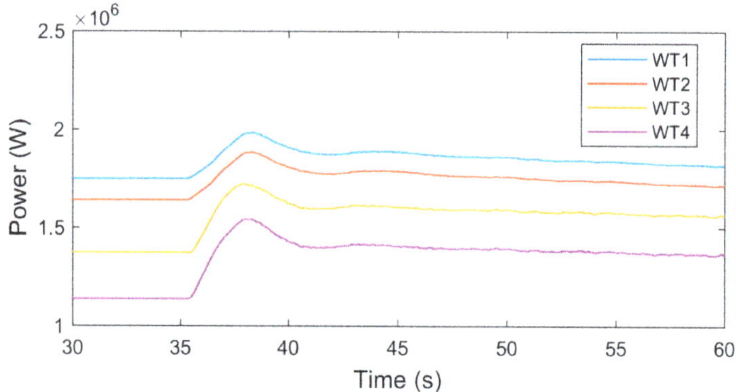

Figure 19. Wind energy systems power (wind farm).

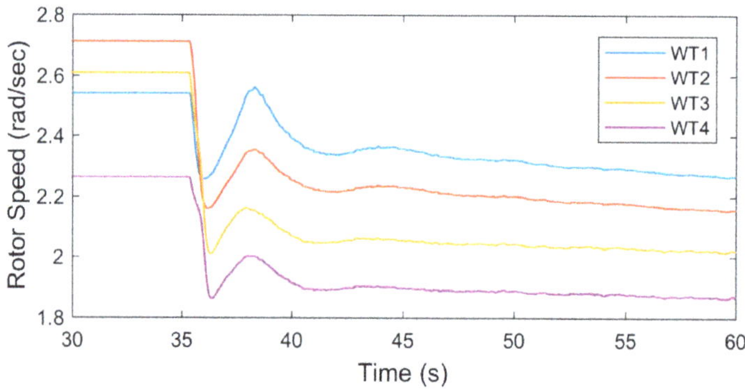

Figure 20. Rotor speed (wind farm).

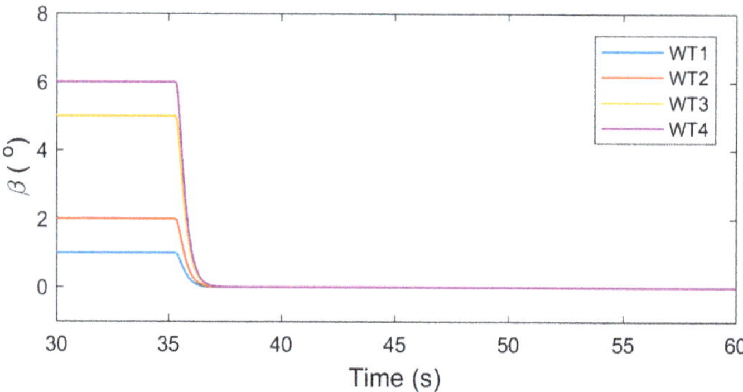

Figure 21. Pitch angle (wind farm).

6. Conclusions

In this article, frequency regulation was provided by a wind farm using a fuzzy logic-based controller. The proposed droop controller maintained stable and proper droop rates while providing adequate frequency support. The droop controller is designed based on the available power of the WES and the rate of change in the grid's frequency at the PCC. The fuzzy logic improved the performance of the dynamic controller. The difference between the nadir point and the nominal value of the frequency is decreased. The constant curve does not work well for WES application because of the variable nature in wind velocity. Even though the fast rate (static) provides a very sharp response against frequency drop, it could lead to instability to the machine-side controller that is responsible for rotational speed. Also, it may increase the stresses and mechanical loading for the structure of the WES. Thus, having a dynamic droop rate that changes based on the availability of wind power and the rate of change of the frequency ensures a stable and smooth response of WES during frequency deviation.

Author Contributions: The authorship is equally shared, where M.G.S. served as the doctoral adviser for A.B. in his Ph.D. program at Colorado School of Mines.

Funding: This research received no external funding.

Conflicts of Interest: The authors declare no conflict of interest.

References

1. Muljadi, E.; Gevorgian, V.; Singh, M.; Santoso, S. Understanding inertial and frequency response of wind power plants. In Proceedings of the 2012 IEEE Power Electronics and Machines in Wind Applications, Denver, CO, USA, 16–18 July 2012; pp. 1–8.
2. Singh, M.; Gevorgian, V.; Muljadi, E.; Ela, E. Variable-speed wind power plant operating with reserve power capability. In Proceedings of the 2013 IEEE Energy Conversion Congress and Exposition, Denver, CO, USA, 15–19 September 2013; pp. 3305–3310.
3. Margaris, I.D.; Papathanassiou, S.; Hatziargyriou, N.D.; Hansen, A.D.; Sorensen, P. Frequency Control in Autonomous Power Systems with High Wind Power Penetration. *IEEE Trans. Sustain. Energy* **2012**, *3*, 189–199. [CrossRef]
4. Pradhan, C.; Bhende, C. Enhancement in Primary Frequency Contribution using Dynamic Deloading of Wind Turbines. *IFAC-PapersOnLine* **2015**, *48*, 13–18. [CrossRef]
5. Camblong, H.; Vechiu, I.; Etxeberria, A.; Martinez, M.I. Wind turbine mechanical stresses reduction and contribution to frequency regulation. *Control Eng. Pract.* **2014**, *30*, 140–149. [CrossRef]
6. Diaz-Gonzalez, F.; Hau, M.; Sumper, A.; Gomis-Bellmunt, O. Coordinated operation of wind turbines and flywheel storage for primary frequency control support. *Int. J. Electr. Power Energy Syst.* **2015**, *68*, 313–326. [CrossRef]
7. Miao, L.; Wen, J.; Xie, H.; Yue, C.; Lee, W. Coordinated Control Strategy of Wind Turbine Generator and Energy Storage Equipment for Frequency Support. *IEEE Trans. Ind. Appl.* **2015**, *51*, 2732–2742. [CrossRef]
8. Van de Vyver, J.; de Kooning, J.D.M.; Meersman, B.; Vandoorn, T.L.; Vandevelde, L. Optimization of constant power control of wind turbines to provide power reserves. In Proceedings of the 2013 48th International Universities' Power Engineering Conference (UPEC), Dublin, Ireland, 2–5 September 2013.
9. El Mokadem, M.; Courtecuisse, V.; Saudemont, C.; Robyns, B.; Deuse, J. Experimental study of variable speed wind generator contribution to primary frequency control. *Renew. Energy* **2009**, *34*, 833–844. [CrossRef]
10. Singarao, V.Y.; Rao, V.S. Frequency responsive services by wind generation resources in United States. *Renew. Sustain. Energy Rev.* **2016**, *55*, 1097–1108. [CrossRef]
11. Moutis, P.; Papathanassiou, S.A.; Hatziargyriou, N.D. Improved load-frequency control contribution of variable speed variable pitch wind generators. *Renew. Energy* **2012**, *48*, 514–523. [CrossRef]
12. Badihi, H.; Zhang, Y.; Hong, H. Active power control design for supporting grid frequency regulation in wind farms. *Annu. Rev. Control* **2015**, *40*, 70–81. [CrossRef]
13. Dreidy, M.; Mokhlis, H.; Mekhilef, S. Inertia response and frequency control techniques for renewable energy sources: A review. *Renew. Sustain. Energy Rev.* **2017**, *69*, 144–155. [CrossRef]
14. Žertek, A.; Member, S.; Verbi, G.; Member, S.; Pantoš, M. A Novel Strategy for Variable-Speed Wind Turbines ' Participation in Primary Frequency Control. *IEEE Trans. Sustain. Energy* **2012**, *3*, 791–799. [CrossRef]
15. Erlich, I.; Wilch, M. Primary frequency control by wind turbines. In Proceedings of the IEEE PES General Meeting, Providence, RI, USA, 25–29 July 2010.
16. Camblong, H.; Nourdine, S.; Vechiu, I.; Tapia, G. Control of wind turbines for fatigue loads reduction and contribution to the grid primary frequency regulation. *Energy* **2012**, *48*, 284–291. [CrossRef]
17. Satpathy, G.; Mehta, A.K.; Kumar, R.; Baredar, P. An overview of various frequency regulation strategies of grid connected and stand-alone wind energy conversion system. In Proceedings of the International Conference on Recent Advances and Innovations in Engineering (ICRAIE-2014), Jaipur, India, 9–11 May 2014.

18. Motamed, B.; Chen, P.; Persson, M. Comparison of primary frequency support methods for wind turbines. In Proceedings of the 2013 IEEE Grenoble Conference, Grenoble, France, 16–20 June 2013.

19. Wang, Y.; Bayem, H.; Giralt-devant, M.; Silva, V.; Guillaud, X.; Francois, B. Methods for Assessing Available Wind Primary Power Reserve. *IEEE Trans. Sustain. Energy* **2015**, *6*, 272–280. [CrossRef]

20. Qian, D.; Tong, S.; Liu, H.; Liu, X. Load frequency control by neural-network-based integral sliding mode for nonlinear power systems with wind turbines. *Neurocomputing* **2016**, *173*, 875–885. [CrossRef]

21. Tielens, P.; de Rijcke, S.; Srivastava, K.; Reza, M.; Marinopoulos, A.; Driesen, J. Frequency support by wind power plants in isolated grids with varying generation mix. In Proceedings of the 2012 IEEE Power and Energy Society General Meeting, San Diego, CA, USA, 22–26 July 2012.

22. Díaz, G.; González-Morán, C.; Gómez-Aleixandre, J.; Diez, A. Scheduling of droop coefficients for frequency and voltage regulation in isolated microgrids. *IEEE Trans. Power Syst.* **2010**, *25*, 489–496. [CrossRef]

23. Molina-Garcia, A.; Munoz-Benavente, I.; Hansen, A.D.; Gomez-Lazaro, E. Demand-Side Contribution to Primary Frequency Control With Wind Farm Auxiliary Control. *IEEE Trans. Power Syst.* **2014**, *29*, 2391–2399. [CrossRef]

24. Aho, J.; Pao, L.Y.; Fleming, P.; Ela, E. Controlling Wind Turbines for Secondary Frequency Regulation: An Analysis of AGC Capabilities under New Performance Based Compensation Policy. Presented at the 13th International Workshop on Large-ScaleIntegration of Wind Power Into Power Systems as Well as onTransmission Networks for Offshore Wind Power Plants, Berlin, Germany, 11–13 November 2014.

25. Aho, J.; Buckspan, A.; Laks, J.H. A tutorial of wind turbine control for supporting grid frequency through active power control. In Proceedings of the 2012 American Control Conference (ACC), Montreal, QC, Canada, 27–29 June 2012.

26. Ela, E.; Tuohy, A.; Milligan, M.; Kirby, B.; Brooks, D. Alternative Approaches for Incentivizing the Frequency Responsive Reserve Ancillary Service. *Electr. J.* **2012**, *25*, 88–102. [CrossRef]

27. Singhvi, V.; Brooks, D.; Member, S. Impact of Wind Active Power Control Strategies on Frequency Response of an Interconnection. In Proceedings of the 2013 IEEE Power & Energy Society General Meeting, Vancouver, BC, Canada, 21–25 July 2013.

28. Miller, N.W.; Clark, K.; Shao, M. Frequency responsive wind plant controls: Impacts on grid performance. In Proceedings of the 2011 IEEE Power and Energy Society General Meeting, San Diego, CA, USA, 24–29 July 2011.

29. Ela, E.; Gevorgian, V.; Fleming, P.A.; Zhang, Y.C.; Singh, M.; Muljadi, E.; Scholbrock, A.; Aho, J.; Buckspan, A.; Pao, L.Y.; et al. *Active Power Controls from Wind Power: Bridging the Gaps*; Technical Report, NREL/TP-5D00-60574; National Renewable Energy Laboratory: Golden, CO, USA, 2014.

30. Vidyanandan, K.V.; Senroy, N. Primary Frequency Regulation by Deloaded Wind Turbines Using Variable Droop. *IEEE Trans. Power Syst.* **2013**, *28*, 837–846. [CrossRef]

31. Khaledian, A.; Golkar, M.A. Analysis of droop control method in an autonomous microgrid. *J. Appl. Res. Technol.* **2017**, *15*, 371–377. [CrossRef]

32. Tarnowski, G.C. *Coordinated Frequency Control of Wind Turbines in Power Systems with High Wind Power Penetration*; Technical University of Denmark (DTU): Lyngby, Denmark, 2012.

33. Baccino, F.; Member, S.; Conte, F.; Grillo, S.; Massucco, S.; Member, S.; Silvestro, F. An Optimal Model-Based Control Technique to Improve Wind Farm Participation to Frequency Regulation. *IEEE Trans. Sustain. Energy* **2015**, *6*, 993–1003. [CrossRef]

34. Buckspan, A.; Aho, J.; Fleming, P.; Jeong, Y.; Pao, L. Combining droop curve concepts with control systems for wind turbine active power control. In Proceedings of the 2012 IEEE Power Electronics and Machines in Wind Applications, Denver, CO, USA, 16–18 July 2012.

35. Aho, J.; Buckspan, A.; Pao, L.Y.; Fleming, P.A. An Active Power Control System for Wind Turbines Capable of Primary and Secondary Frequency Control for Supporting Grid Reliability. In Proceedings of the 51st AIAA Aerospace Sciences Meeting including the New Horizons Forum and Aerospace Exposition, Aerospace Sciences Meetings, Grapevine, TX, USA, 7–10 January 2013.

36. Delille, G.; François, B.; Malarange, G. Dynamic frequency control support by energy storage to reduce the impact of wind and solar generation on isolated power system's inertia. *IEEE Trans. Sustain. Energy* **2012**, *3*, 931–939. [CrossRef]

37. Guo, X.Q.; Lu, Z.G.; Wang, B.C.; Sun, X.F.; Wang, L.; Guerrero, J.M. Dynamic Phasors-Based Modeling and Stability Analysis of Droop-Controlled Inverters for Microgrid Applications. *IEEE Trans. Smart Grid* **2014**, *5*, 2980–2987. [CrossRef]
38. Arani, M.F.M.; Mohamed, Y.A.-R.I. Dynamic Droop Control for Wind Turbines Participating in Primary Frequency Regulation in Microgrids. *IEEE Trans. Smart Grid* **2018**, *9*, 5742–5751. [CrossRef]
39. Wang, Y.; Chen, Z.; Deng, F. Dynamic droop scheme considering effect of intermittent renewable energy source. In Proceedings of the 2016 IEEE 7th International Symposium on Power Electronics for Distributed Generation Systems (PEDG), Vancouver, BC, Canada, 27–30 June 2016.
40. Hwang, M.; Chun, Y.; Park, J.; Kang, Y.C. Dynamic Droop-based Inertial Control of a Wind Power Plant. *J. Electr. Eng. Technol.* **2015**, *10*, 1363–1369. [CrossRef]
41. Kundur, P. *Power System Stability and Control*; McGraw-Hill: New York, NY, USA, 1995.
42. Anderson, P.M.; Fouad, A.A. *Power System Control and Stability*; John Wiley & Sons, Ltd.: Hoboken, NJ, USA, 1999.
43. Sun, Y.; Zhang, Z.; Li, G.; Lin, J. Review on frequency control of power systems with wind power penetration. *Power Syst. Technol.* **2010**, 1–8. [CrossRef]
44. Janssens, N.A.; Member, S.; Lambin, G.; Bragard, N. Active Power Control Strategies of DFIG Wind Turbines. In Proceedings of the 2007 IEEE Lausanne Power Tech, Lausanne, Switzerland, 1–5 July 2007.
45. Zadeh, L.A. Outline of a new approach to the analysis of complex systems and decision processes. *Syst. Man Cybern. IEEE Trans.* **1973**, *SMC-3*, 28–44. [CrossRef]
46. Muyeen, S.M.; Al-Durra, A. Modeling and control strategies of fuzzy logic controlled inverter system for grid interconnected variable speed wind generator. *IEEE Syst. J.* **2013**, *7*, 817–824. [CrossRef]
47. Ahmadi, S.; Shokoohi, S.; Bevrani, H. A fuzzy logic-based droop control for simultaneous voltage and frequency regulation in an AC microgrid. *Int. J. Electr. Power Energy Syst.* **2015**, *64*, 148–155. [CrossRef]
48. Putri, A.I.; Ahn, M.; Choi, J. Speed sensorless fuzzy MPPT control of grid-connected PMSG for wind power generation. In Proceedings of the 2012 International Conference on Renewable Energy Research and Applications (ICRERA), Nagasaki, Japan, 11–14 November 2012.
49. Hoseinzadeh, B.; Chen, Z. Intelligent load-frequency control contribution of wind turbine in power system stability. In Proceedings of the Eurocon 2013, Zagreb, Croatia, 1–4 July 2013; pp. 1124–1128.
50. Lagorse, J.; Simões, M.G.; Miraoui, A. A Multiagent Fuzzy-Logic-Based Energy Management of Hybrid Systems. *IEEE Trans. Ind. Appl.* **2009**, *45*, 2123–2129. [CrossRef]
51. Bubshait, A.S.; Simões, M.G.; Mortezaei, A.; Busarello, T.D.C. Power quality achievement using grid connected converter of wind turbine system. In Proceedings of the 2015 IEEE Industry Applications Society Annual Meeting, Addison, TX, USA, 18–22 October 2015.
52. Bubshait, A.; Simões, M.G. Optimal Power Reserve of a Wind Turbine System Participating in Primary Frequency Control. *Appl. Sci.* **2018**, *8*, 2022. [CrossRef]

Article

Reactive Power Optimization for Transient Voltage Stability in Energy Internet via Deep Reinforcement Learning Approach

Junwei Cao, Wanlu Zhang, Zeqing Xiao and Haochen Hua *

Research Institute of Information Technology, Beijing National Research Center for Information Science and Technology, Tsinghua University, Beijing 100084, China; jcao@tsinghua.edu.cn (J.C.); zhangwl15@tsinghua.org.cn (W.Z.); xiaozeqing@mail.tsinghua.edu.cn (Z.X.)
* Correspondence: hhua@tsinghua.edu.cn

Received: 12 March 2019; Accepted: 22 April 2019; Published: 24 April 2019

Abstract: The existence of high proportional distributed energy resources in energy Internet (EI) scenarios has a strong impact on the power supply-demand balance of the EI system. Decision-making optimization research that focuses on the transient voltage stability is of great significance for maintaining effective and safe operation of the EI. Within a typical EI scenario, this paper conducts a study of transient voltage stability analysis based on convolutional neural networks. Based on the judgment of transient voltage stability, a reactive power compensation decision optimization algorithm via deep reinforcement learning approach is proposed. In this sense, the following targets are achieved: the efficiency of decision-making is greatly improved, risks are identified in advance, and decisions are made in time. Simulations show the effectiveness of our proposed method.

Keywords: energy Internet; convolutional neural network; decision optimization; deep reinforcement learning

---

1. Introduction

With the development of renewable energy related technology, our dependence on conventional energy has been gradually declining. As the core of the third industrial revolution, a new concept named as the energy Internet (EI) has been proposed and investigated extensively [1,2], in which a new architecture of energy supply and demand is constructed through the integration of information and energy [3–5]. Typically, an EI scenario can have access to the utility grid. Alternatively, when disconnected from the main power grid, multiple sub-grids interconnected via energy routers are able to function normally [6,7]. For the detailed definition, architecture and key technologies of EI, readers can refer to [8,9], and the references therein.

Due to the increase of uncertainty in power generation and usage, compared with traditional power grids, one of the challenges faced by EI is how to match power demand with supply and how to maintain the safety and reliability of the whole network. The problem of resilient multi-scale coordination control against a set of adversarial or non-cooperative nodes in directed networks has been investigated in [10]. In power systems, static and transient voltage stability analysis have been extensively studied; see, e.g., [11]. Transient voltage stability problems, such as voltage sag, may occur in a local network that is not robust in the event of a large disturbance. It is notable that such transient voltage stability issues also exist in the field of EI, which is worth investigation [12].

The loss of reactive power can increase the voltage loss and may also lead to voltage fluctuation. Reactive power compensation is of great significance for the safe and reliable operation of EI. The following four targets can be achieved by proper reactive power compensation: (1) stabilizing the grid voltage, (2) increasing the power factor, (3) improving the equipment utilization rate, (4) reducing the loss of network active power; see, e.g., [13]. In order to guarantee the normal operation of EI, the dilemma caused by reactive power consumption can be solved by installing a static var generator (SVG) [14]. The SVG, also known as STATCOM, is a commonly used device to solve the reactive power consumption problem [15]. The work principle of SVG is as follows: the voltage source inverter is connected in parallel to EI. Amplitude and phase of output voltage on the AC side is adjusted, or the current on the AC side is directly adjusted to absorb or emit reactive power. Thus, reactive power compensation can be dynamically implemented. On the basis of stability assessment and prediction, SVG is installed to maintain the safe and stable operation of EI. For the case that voltage stability is not restored by the self-healing ability of EI, the installation of SVGs at different locations and the setting of different SVGs' output reactive power affects the voltage stability and the time of restoring stability. For the case of restoring voltage stability through the self-healing ability of EI, the restoration speed can be accelerated by installing the SVG. Additionally, the influence on the power consumption on the customer side can be reduced [16].

Within EI scenarios, transient short-circuit failure may cause great economic loss [17]. The judgment of transient voltage stability is not only the basis for subsequent decision optimization of reactive power compensation, but also the key to maintaining the normal operation of EI. Additionally, the credibility of subsequent decision optimization is affected by the accuracy of the judgment of the stability state. At present, the mainstream conventional methods used for the judgement of the transient voltage stability state are mainly time domain simulation approaches [18] and direct methods [19], which are based on deterministic analysis. Due to the intermittence and volatility of power generation by renewable energy sources, judgement of the transient voltage stability state cannot be analyzed via deterministic approaches.

In recent years, with the development of big data technology and data mining technology, machine learning algorithms have been applied to the judgement of the transient voltage stability state [20]. The aforementioned algorithms mainly include artificial neural network, decision tree, support vector machine (SVM) [21] and other shallow machine learning algorithms. To illustrate, an intelligent algorithm using forward feedback neural network for online voltage stability assessment and monitoring has been studied in [22], where voltage, active power, reactive power of generators and loads are used as characteristic inputs for online voltage stability evaluation. In [21], the SVM algorithm was applied to select the voltage level, generator rate and rotor angle as input features for the prediction and evaluation of transient voltage stability after any fault occurs. In [23], the authors propose a voltage safety evaluation method through regularly updating the decision tree. The multi-layer perceptron neural network is employed to select new characteristics of voltage value and reactive power generation for online voltage stability testing and evaluation [24]. In [25], the extreme learning machine is used for voltage stability margin evaluation.

It is notable that the rapidity and accuracy of optimization is difficult to be achieved simultaneously by conventional evaluation methods. The shallow machine learning algorithm that processes the input characteristics of complex classification problems has limited computing power, which cannot meet the accuracy requirement of transient voltage stability prediction and evaluation in EI. In recent years, extensive applications of deep learning in the field of transient voltage stability prediction and evaluation have been used to solve the aforementioned challenge. Deep learning has a strong feature extraction ability and can solve dimensional disaster problems including multi-nodes and multi-features in EI; see, e.g., [26]. At present, the commonly used deep learning algorithms include the deep belief network [27,28], recurrent neural network [29], stacked denoising auto-encoders [30] and convolutional neural network (CNN) [31]. The combination of deep learning and reinforcement learning forms the deep reinforcement learning approach [32,33]. Reinforcement learning can be viewed as a process of exploration in the unknown environment [34]. From environment mapping to action, the subject not only obtains the action with the maximum reward value through exploration, but also receives the ultimate optimal effect by continuous trials and errors. Thus, as the ultimate goal, the maximum cumulative reward value is obtained. Reinforcement learning mainly includes four key aspects: strategy, reward, evaluation and environment. Such methods explore the unknown environment, and different strategic selection actions are performed to obtain different reward and punishment values, such that the quality of the strategy is evaluated. It is worth mentioning that the quality of the evaluation goal is limited to the reward value obtained after the completion of an action. Besides, it depends on the follow-up action and the reward value obtained eventually. Based on the environment of discrete-time Markov decision process, the Q learning algorithm is one of the most important algorithms in reinforcement learning [35].

The data in EI include information about the state of each node at each time point and information about the network topology, and such data has both time and spatial correlation. The conventional simplified power network model [36] based on simulation fails to make full use of the real-time information obtained by massive data acquisition devices. In addition, the decision-making on reactive power compensation in existing power grids is mainly based on manual operations. In this paper, reactive power optimization for transient voltage stability in EI is studied. Based on data with sufficient information, a deep reinforcement learning model is used to judge the transient voltage stability state. In order to avoid losing information about time and space while training the model, CNN is selected to predict the transient voltage stability. Next, based on the stability prediction results, the deep reinforcement learning algorithm is applied to the decision optimization of reactive power compensation. Simulations show the effectiveness of the proposed method.

The contribution of this paper can be outlined as follows:

(1) A judgement model for stability state of EI based on a deep learning algorithm is proposed. Compared to the conventional simplified power network models [36], this paper proposes a data-based method for reactive power optimization, such that transient voltage stability is achieved. Compared to a model-based method, the error caused by a data-based method is smaller. In this sense, the efficiency of data processing has been improved. Therefore, the efficiency of decision-making of reactive power compensation is greatly improved, such that the desired reliable operation of EI can be achieved.

(2) By analyzing the data in each node of the EI within the feature time period, the deep learning algorithm is used to train the stability judgement model. In this sense, whether the voltage would return to a stable state or not after the short-circuit failure occurs can be estimated, which provides delay stability information for decision optimization. Meanwhile, based on the mainstream power simulation software Bonneville Power Administration (BPA) [37], the data batch processing toolkit is developed to facilitate the change of data card in batches and to extract data in batches.

(3) Traditional power grid decision-making is mainly based on the experience of grid operators and manual operation. In this paper, an advanced artificial intelligence-based method is applied to the decision-making optimization of reactive power compensation in EI. In this sense, the efficiency of decision-making and the accuracy of optimization has been greatly improved.

The rest of this paper is organized as follows. Section 2 provides the problem formulation. Section 3 constructs the flow of reactive power decision optimization algorithm. Section 4 provides some simulations. Finally, we conclude our paper in Section 5.

2. Problem Formulation

The goal of this paper is to achieve the stability of high voltage buses, complete distributed reactive power compensation, and minimize the total compensation of the SVG. In this paper, for the training model, deep learning and reinforcement learning are combined to provide the installation strategy of SVGs.

2.1. Voltage Stability in EI and Construction of the Simulation Model

The local microgrids can be integrated into the large power grid, or they can be operated in the islanded mode. When the local-area grid network is disconnected from the large power grid, voltage stability issues occur, which potentially affects the reliability of the system operation.

Normally, an EI scenario functions in a stable state for most of the operation time, and an unstable state caused by short-circuit faults rarely occurs. Thereby, the gaps between the number of stable and unstable samples collected by phasor measurement unit (PMU) in EI are extremely large. If real data is used for prediction and all the selected classifiers are stable, the accuracy of the trained classifier would still be relatively high (no less than 99%). In this manner, there is no training effect. Hence, for the considered EI scenario in this paper, the simulation data is generated by BPA software. The power grid simulation model is shown in Figure 1. The sequence numbers 1–60 in Figure 1 represent network nodes 1–60. We consider n voltage-grade substations, namely, n_i kV, and A_i substation, $1 \leq i \leq n$. Here, A_i is a custom symbol.

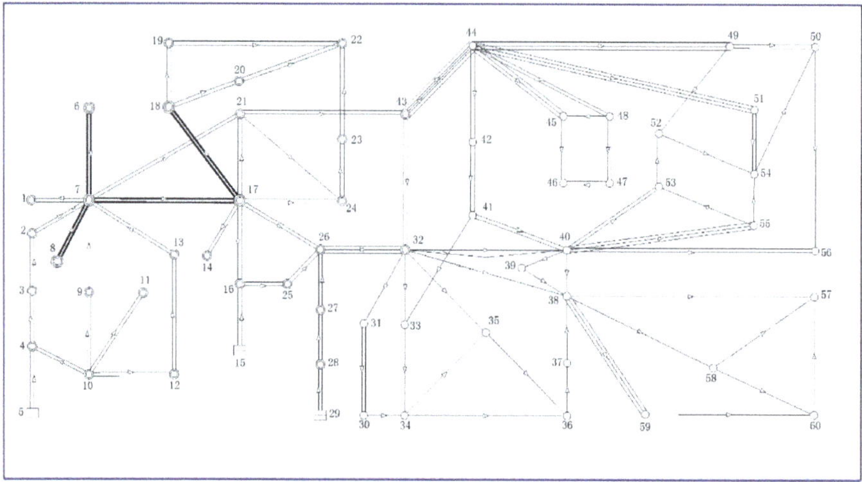

Figure 1. Energy Internet (EI) simulation model.

2.2. Judgment of Transient Voltage Stability

The data extraction program file was written, and the data in each BPA output file was extracted. The voltage U, frequency f, active power P and reactive power Q of each node measured each half cycle in the first n cycles is taken as the input data in the process of training the stability evaluation model. The data of voltage U in the last five cycles is taken to determine whether the value of voltage in the stable state is finally restored. The judgment result is used as the output data in the process of training the stability evaluation model.

The results of the stability prediction are evaluated by three indexes: precision, recall, and f1-score (known as the harmonic mean of precision and recall [38]), which are as follows:

$$precision = \frac{true\ positive}{true\ positive + false\ positive}$$

$$recall = \frac{true\ positive}{true\ positive + false\ negative}$$

$$f1 - score = 2 \times \frac{precision \times recall}{precusuib + recall}$$

The interpretations of *true positive*, *false positive*, *false negative* and *true negative* are shown in Table 1. *True* means that the classification is correct, and *false* means that the classification is false. *Positive* means that classification is positive sample "1", and *negative* means that classification is negative sample "0". By developing a general mathematical framework based upon the percolation model, [39] investigates attack robustness analytically with a false positive/negative rate.

Table 1. Evaluation of classification results.

Predicted Value	Actual Value	
	1	**0**
1	*true positive*	*false positive*
0	*false negative*	*true negative*

3. Optimization Algorithm

In this section, we introduce the flow of reactive power decision optimization algorithm based on the judgement of the transient voltage stability state.

3.1. The Algorithm for Judging the Transient Voltage Stability State Based on CNN

For the prediction of transient voltage stability, normally, the selected input characteristics are only time domain data. Single-node historical data is mainly considered, without taking into account the overall spatial characteristics of the grid. Therefore, the historical data of other nodes which contain a large amount of valid information that is useful for the stability prediction of such nodes is missed. In addition, the massive collected PMU data fails to be properly processed. In this paper, the judgment algorithm of the transient voltage stability state is designed based on the analysis of each node's data acquired by PMU. Meanwhile, the distance between the time period of the selected characteristic and the time to be predicted is enlarged. In this sense, the prediction effect can be achieved in advance.

The detailed algorithm for judging the transient voltage stability is as follows:

Step 1: Establishment of the model input sample matrix.

Real-time data is acquired from the data acquisition device PMU deployed at each key node of the real power grid. The output data can be obtained by simulation software. The data values of voltage U, frequency f, active power P and reactive power Q of the key nodes in EI during a characteristic time period T are obtained. The input sample matrix that makes up the model is as follows:

$$\begin{bmatrix} U_{1,1} & f_{1,1} & P_{1,1} & Q_{1,1} & \cdots & U_{1,M} & f_{1,M} & P_{1,M} & Q_{1,M} \\ U_{2,1} & f_{2,1} & P_{2,1} & Q_{2,1} & \cdots & U_{2,M} & f_{2,M} & P_{2,M} & Q_{2,M} \\ U_{3,1} & f_{3,1} & P_{3,1} & Q_{3,1} & \cdots & U_{3,M} & f_{3,M} & P_{3,M} & Q_{3,M} \\ \vdots & \vdots & \vdots & \vdots & \ddots & \vdots & \vdots & \vdots & \vdots \\ U_{N,1} & f_{N,1} & P_{N,1} & Q_{N,1} & \cdots & U_{N,M} & f_{N,M} & P_{N,M} & Q_{N,M} \end{bmatrix}$$

where the subscript of voltage U, frequency f, active power P and reactive power Q is (i, j). The first subscript i represents the i-th sample. The second subscript j represents the j-th time collection point.

Step 2: Determination and labeling of the input data stability.

According to the industrial standard, the stability of the input sample data is labeled. The value of voltage U at a specific time is used to determine whether the voltage is stable or not. If the value of node voltage U returns to 0.8 times of the standard value, it is regarded as stable and is denoted as "1". Conversely, if it is considered as unstable, it is denoted as "0".

Step 3: Expansion of data.

Considering the imbalance of positive and negative samples under the situation of stability and instability, the input sample data is expanded by translating window, in order to avoid deflection in the training process. Such process is illustrated in Figure 2.

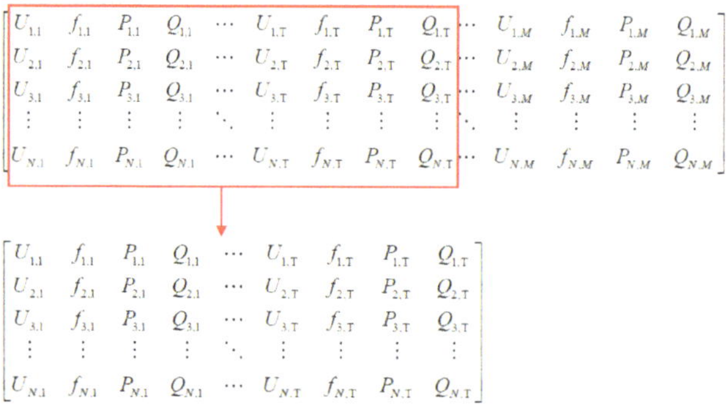

$$\begin{bmatrix} U_{1,1} & f_{1,1} & P_{1,1} & Q_{1,1} & \cdots & U_{1,T} & f_{1,T} & P_{1,T} & Q_{1,T} & \cdots & U_{1,M} & f_{1,M} & P_{1,M} & Q_{1,M} \\ U_{2,1} & f_{2,1} & P_{2,1} & Q_{2,1} & \cdots & U_{2,T} & f_{2,T} & P_{2,T} & Q_{2,T} & \cdots & U_{2,M} & f_{2,M} & P_{2,M} & Q_{2,M} \\ U_{3,1} & f_{3,1} & P_{3,1} & Q_{3,1} & \cdots & U_{3,T} & f_{3,T} & P_{3,T} & Q_{3,T} & \cdots & U_{3,M} & f_{3,M} & P_{3,M} & Q_{3,M} \\ \vdots & \vdots & \vdots & \vdots & \ddots & \vdots & \vdots & \vdots & \vdots & \ddots & \vdots & \vdots & \vdots & \vdots \\ U_{N,1} & f_{N,1} & P_{N,1} & Q_{N,1} & \cdots & U_{N,T} & f_{N,T} & P_{N,T} & Q_{N,T} & \cdots & U_{N,M} & f_{N,M} & P_{N,M} & Q_{N,M} \end{bmatrix}$$

$$\begin{bmatrix} U_{1,1} & f_{1,1} & P_{1,1} & Q_{1,1} & \cdots & U_{1,T} & f_{1,T} & P_{1,T} & Q_{1,T} \\ U_{2,1} & f_{2,1} & P_{2,1} & Q_{2,1} & \cdots & U_{2,T} & f_{2,T} & P_{2,T} & Q_{2,T} \\ U_{3,1} & f_{3,1} & P_{3,1} & Q_{3,1} & \cdots & U_{3,T} & f_{3,T} & P_{3,T} & Q_{3,T} \\ \vdots & \vdots & \vdots & \vdots & \ddots & \vdots & \vdots & \vdots & \vdots \\ U_{N,1} & f_{N,1} & P_{N,1} & Q_{N,1} & \cdots & U_{N,T} & f_{N,T} & P_{N,T} & Q_{N,T} \end{bmatrix}$$

Figure 2. Data expansion mode in the case of unbalanced samples.

Step 4: Construction of CNN.

The CNN is constructed by input layer, convolution layer, pooling layer, fully connected layer and output layer. The appropriate number of CNN layers, convolutional cores and parameters are selected to achieve a better prediction effect.

Step 5: Offline training and online evaluation.

The combination model of offline training and online evaluation is shown in Figure 3. According to the transient stability rule, the transient stability assessment model is obtained by offline training using historical data or simulation data. Then, real time data is used in the trained model for online testing, and the stability assessment results are obtained.

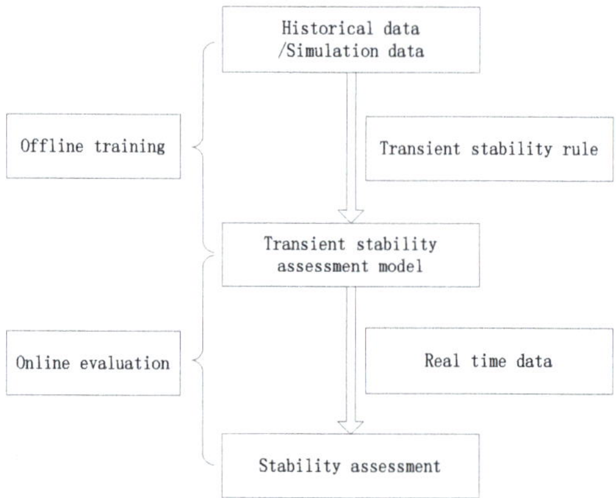

Figure 3. Combination model of offline training and online evaluation.

3.2. Reactive Power Decision Optimization Algorithm

Based on the judgement of the transient voltage stability state, the process of the reactive power decision optimization algorithm is proposed as follows:

Step 1: State perception.

The output data is obtained through BPA. During a characteristic time period T, voltage U, frequency f, active power P and reactive power Q of each key node are selected to form an input sample matrix of the model as the current state s.

Step 2: Stability prediction.

According to the judgement in Section 3.1, the state information perceived in Step 1 is taken as the input data of the model, and the output is whether the grid would restore stability or not within a certain time period. The stable output is used as an important basis for calculating the reward value by the subsequent deep reinforcement learning approach.

Step 3: Capture of action.

The location of SVGs and compensation value of each SVG are used as action a of operator *Agent* in the deep reinforcement learning algorithm. The action value is acquired according to the setting mode in the effective action collection and then converted into a *one − hot* form.

Step 4: Perception of the next state.

In the case of the perceived state s in Step 1, the position and compensation value of SVGs are set in BPA by executing action a obtained in Step 3. The next state value s, is obtained by performing the simulation.

Step 5: Reward value setting.

There are two goals for reactive power optimization. The first one is to enable the grid to recover in a certain time period after a short-circuit fault occurs. The other is to use distributed reactive power compensation, so as to reduce the compensation value of each reactive power compensator. The calculation rule of the reward value r is set in conjunction with the stability prediction in Step 2. The action value is acquired in Step 3.

Step 6: Experiential playback.

The collected status, action, reward and other data are stored in the database *memory_replay* which is self-defined. The training data is randomly selected in small batches during training. In this sense, the dependency relationship of the observed data can be avoided. In addition, the effect of the operator *Agent* influenced by the recent operation can also be avoided. Otherwise, what happens before would be "forgotten". The correlation of the samples is weakened, the efficiency of data usage can be improved, and the correlation between data can be reduced. Therefore, the algorithm can be easily convergent, and the generalization ability can be improved.

Step 7: Training of Q network.

The CNN is used to fit Q value function. The experiential playback technique in Step 6 is adopted. The small batch is randomly taken from the database *memory_replay* for training.

The goal is to obtain the action combination of the highest Q value and to output this value.

4. Simulation Results

4.1. Experimental Results and Analysis

According to Section 2.2, we select $N = 500$ and $n = 200$. The loss value curve of the experimental results is shown in Figure 4.

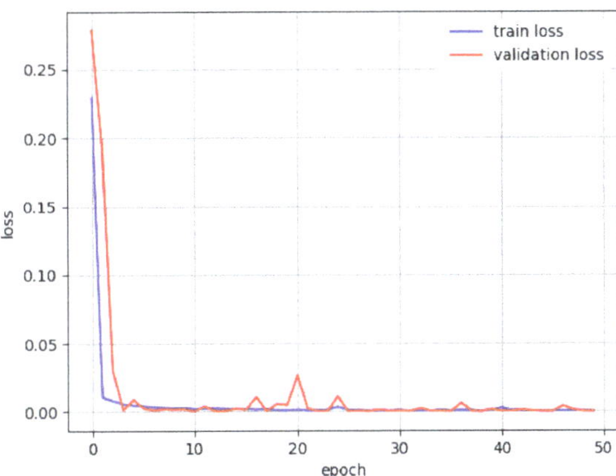

Figure 4. Comparison diagram of loss value of training set and verification set.

The results of the stability prediction on the test set are shown in Figure 5. Taking 0.5 as the dividing line, the value that is greater than 0.5 is classified as "1", which is considered as stable. The value that is less than 0.5 is classified as "0", which is considered as unstable. The accuracy is 2294/2300.

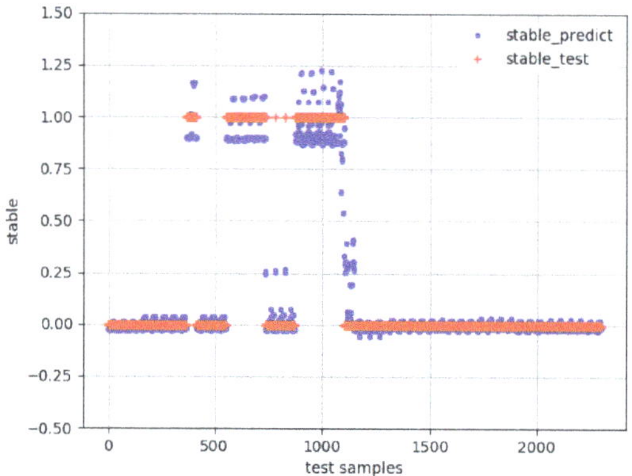

Figure 5. Stability prediction results of test set (first 200 cycles).

It is shown in Table 2 that the false negative value is 6, i.e., six samples are actually stable. However, the prediction result is unstable. In order to maintain the stable operation of EI, it is acceptable to consider appropriate over-warning in real scenarios. While the false positive value is 0, i.e., the actual unstable sample is correctly predicted, which meets our requirements. In this paper, the threshold of accuracy is set to be 99%. The calculation is available with a precision of 1, a recall of 98.7%, and an f1-score of 99.4%.

Table 2. Stability prediction results (first 200 cycles).

Predicted Value	Actual Value	
	Stable	Unstable
Stable	462	0
Unstable	6	1832

The evaluation of stability prediction at this stage is to prepare for the decision optimization of subsequent reactive power compensation. In order to know in advance whether the EI would return to a stable state in the future, the characteristic time period is expected to be reduced. The data from the first 100 cycles would be intercepted, and the aforementioned CNN is still used for training. The accuracy is 2252/2300, which also meets the expected requirements.

The results of stability prediction are specifically analyzed in Table 3. Through calculating, the precision, the recall and the f1-score are 91.2%, 99.4% and 95.1%, respectively. The accuracy does not meet the requirement.

Table 3. Stability prediction results (first 100 cycles).

Predicted Value	Actual Value	
	Stable	Unstable
Stable	465	45
Unstable	3	1787

Although the precision of the prediction has achieved 97.9%, such rate is still relatively low and does not meet the requirement. Therefore, the structure of CNN is determined to be optimized. The parameters in the CNN is set to be fixed. The size of the convolution kernel is selected as 3×3. The depth of the network is adjusted by changing the number of convolution layers. The number of feature extractions is altered by changing the number of convolution kernels per layer, which makes the extraction effect much better. After twenty experiments are conducted, the average value of the training results of different CNN structures with different parameters is obtained. The results are shown in Table 4.

Table 4. Parameter settings of the convolutional neural network (CNN).

Number of Layers	Number of Convolution Kernels	Computation Time (s)	Precision (%)	Recall (%)	f1-Score (%)
4	64	3	91.2	99.4	95.1
4	128	6	95.3	87.4	91.2
4	256	15	95.0	97.9	96.4
6	64	4	99.5	89.3	94.1
6	128	7	95.0	98.3	96.6
6	256	19	98.6	93.4	96.0
8	64	4	94.0	80.8	86.9
8	128	8	94.3	84.4	89.1
8	256	21	94.8	93.6	94.2

It can be seen from Table 4 that as the number of convolution kernels increases, the time for calculation increases substantially. The effect of increasing the number of the convolution layer on the calculation time is not obvious. The setting of the number of convolution layers and convolution kernels affects the prediction results. After twenty experiments are implemented, when the 6-layer convolution is set, and the number of convolution kernels per convolution layer is 64, the duration of calculation becomes shorter, and the accuracy is higher. The false positive value is smaller. More than half of the false positive values in the experimental results are all 0, which implies that the samples that are actually unstable are successfully predicted to be unstable. In this sense, the goal of stability prediction in this paper has been achieved. Thus, the model of CNN is selected. The results of the stability prediction on the test set are shown in Figure 6. The accuracy is 2236/2300, which meets the expected target.

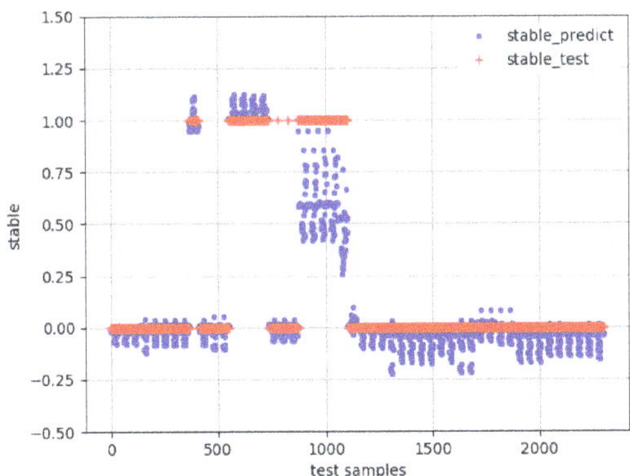

Figure 6. Stability prediction results of the test set after optimization (the first 100 cycles).

The analysis of the stability prediction results is shown in Table 5. The false negative value is 64, i.e., 64 samples are actually stable. However, the prediction result is unstable, which is acceptable in the real engineering scenario. The false positive value is 0, i.e., the actual unstable samples are not predicted to be stable, which meets the expected requirements. Compared with the result before optimization, although the false negative value has increased, such value is reduced to be 0. In real-world applications, a false negative is acceptable, and a false positive is unacceptable. Therefore, the optimized results appear to meet the expected requirements. Through computation, the precision, the recall and the f1-score are 1, 86.3% and 92.7%, respectively.

Table 5. Stability prediction results after optimization (first 100 cycles).

Predicted Value	Actual Value	
	Stable	Unstable
Stable	404	0
Unstable	64	1832

The output of existing research on stability prediction is mainly based on the analysis of single node data. Differently, this study selects the data of each node of the whole network as input data. Meanwhile, the data format is different. In order to verify the feasibility and effectiveness of the study as a comparative experiment, the SVM algorithm [16] is used for training. The false negative value is 141, i.e., 141 samples are stable. However, the prediction result is unstable, which is acceptable. The false positive value is 25, i.e., 25 samples are unstable but these samples are predicted to be stable. In real engineering practice, important information about voltage instability would be missed in such results. The future unstable states are difficult to be predicted accurately. Therefore, it is difficult to make a judgement, which is unacceptable. The obtained precision, the recall and the f1-score are 92.9%, 69.9% and 79.7%, respectively, which are much lower than the result obtained by our algorithm.

In summary, when data for each node in the whole grid system is used as high-dimensional input feature data within a certain characteristic time period, feature extraction can be better performed by deep learning CNN than by the conventional machine learning algorithm. A satisfactory fitting effect is obtained, and the application result in transient voltage stability judgment of EI achieves the expected target. Based on the mainstream power simulation software, a data batch processing toolkit has been developed, which improves the efficiency of data processing.

4.2. Simulation Example of Reactive Power Decision Optimization

BPA is used to simulate different short-circuit faults. Since only the load model and load rate can be changed in BPA, the real-time load value cannot be collected. The system recovery time (cycle) corresponding to different load rates and the total SVG compensation is shown in Figure 7.

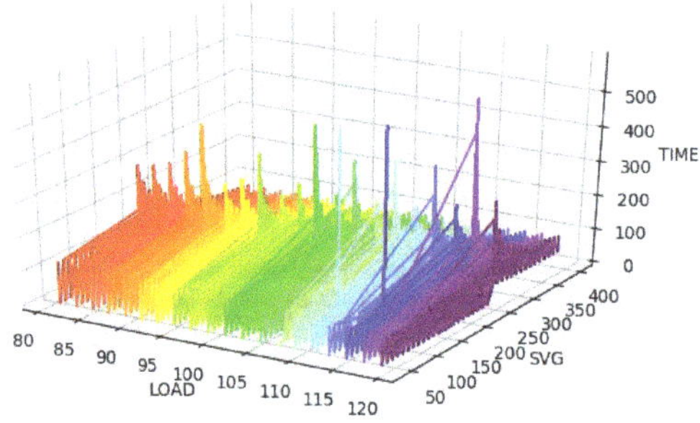

Figure 7. The curve of load rate, SVG compensation, and time of returning to a stable state.

It can be seen that under the same load rate, as the amount of SVG compensation increases, the system fluctuates from an unstable state to a stable state, and the time of returning to the stable state decreases gradually.

One set of SVG compensation was extracted. The load rate-time of returning to stable state (cycle) curve is drawn under the same SVG compensation. As is shown in Figure 8, with the increase in the load rate, the time of returning to a stable state increases gradually after the same SVG compensation is obtained until stability cannot be restored. In Figure 6, the system instability is represented by the time of returning to a stable state of 0.

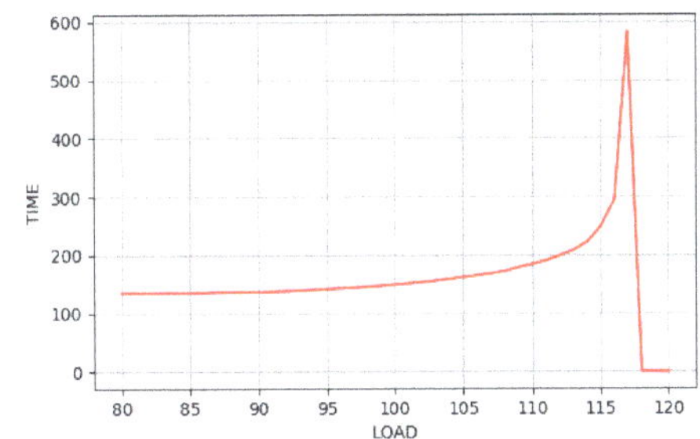

Figure 8. The curve of load rate–time of returning to a stable state when the SVG compensator is the same.

A numerical example is given in the following study where site *A* and site *B* represent for two cities. The load model is set as follows: (1) 30% constant resistance load, 40% constant current load and 30% constant power load at site *A*; and (2) 70% motor and 30% constant impedance with a load rate of 115% at site *B*. The short-circuit fault of single-circuit three-phase is set on buses with different voltage grades. Eight faults are considered in this example, including fault 1 (high-voltage level bus fault of 220 kV between site *A* and site *B*), fault 2 and fault 3 (220 kV high voltage grade bus fault on site *B*, i.e., the grid location studied in the experiment), as well as fault 4 to fault 8 (low voltage bus fault of 110 kV and 66 kV on site *B*).The fault clearing time is taken as 5 cycles, i.e., 0.1 s. Five SVGs are set at two 220 kV high voltage grade substations (A1 substation and A2 substation) and three 110 kV voltage grade substations (A3 substation, A4 substation and A5 substation), respectively. The compensation values of five SVGs, which are intervals of the action in the proposed algorithm, are set as follows: [20 Mvar, 80 Mvar], [50 Mvar, 80 Mvar], [60 Mvar, 80 Mvar], [50 Mvar, 80 Mvar] and [50 Mvar, 80 Mvar]. In the proposed algorithm, the action space is discretized, and the discretized step is 10 Mvar. By this means, 1344 discrete actions are obtained. These actions are numbered from 1 to 1344. These actions are further converted into *one − hot* form. The network is trained by CNN. The loss value is shown in Figure 9.

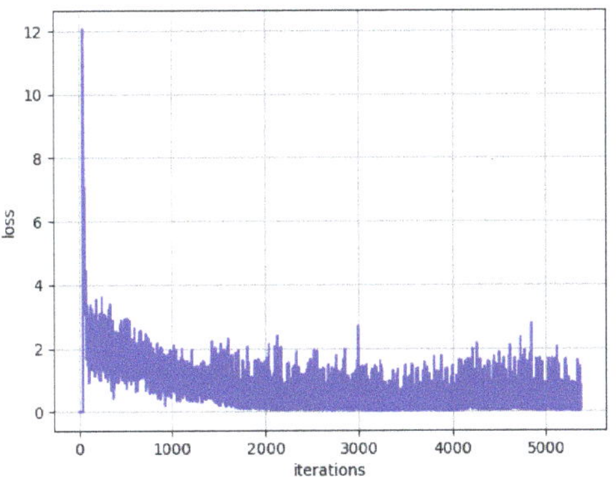

Figure 9. Trend graph of loss value.

The optimization results based on deep reinforcement learning are shown in Table 6. The compensations value of each action in Table 5 for five SVGs are listed in Table 7. These two tables can be understood as follows. Taking the number in the first row of Table 6 as an example, when the network training times are 5000, the output action numbers of fault 1 to fault 8 are 34. According to the second line of Table 7, action 34 compensates for five SVGs with 20 Mvar, 50 Mvar, 80 Mvar, 50 Mvar and 60 Mvar, respectively.

Table 6. Decision Optimization Results.

Training Times	Number of Output Action
5000	34, 34, 34, 34, 34, 34, 34, 34
10,000	105, 105, 105, 105, 105, 105, 105, 105
15,000	389, 389, 389, 389, 389, 389, 389, 389
20,000	201, 201, 201, 241, 241, 241, 241, 241
25,000	19, 19, 19, 241, 241, 241, 241, 241

Table 7. Compensation Values.

Action	Compensation Value (Mvar)
19	20, 50, 70, 50, 70
34	20, 50, 80, 50, 60
105	20, 70, 60, 70, 50
201	30, 50, 60, 70, 50
241	30, 60, 60, 50, 50
389	40, 50, 60, 60, 50

Fault 1, fault 2 and fault 3 are in the high voltage bus. Thus, reactive compensation should be increased. It can be seen that at the initial stage of training, the differences from fault 1 to fault 3 and from fault 4 to fault 8 are not successfully identified. Similar action outputs with high compensation are given. However, the decision of reactive compensation for different grades could be given by increasing the training times. The difference can be seen in the results of later training.

The contrast curve of Q value and reward value are shown in Figure 10. It can be seen that the general trend of the Q value is consistent with the general trend of the reward value. As the training time increases, the Q value is constantly close to the reward value, which achieves the goal of training.

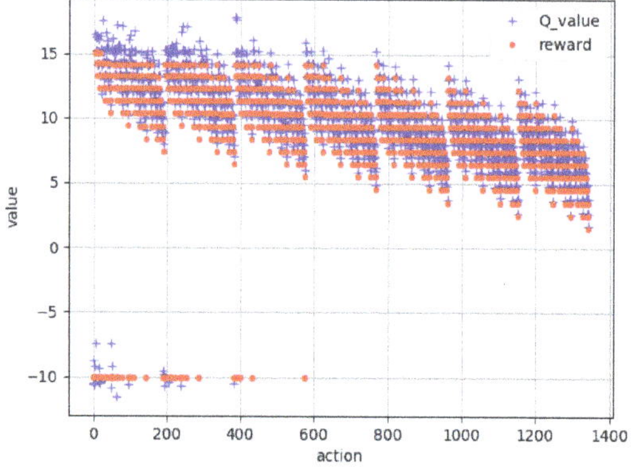

Figure 10. Comparison chart of Q value and reward value (25000 times of training).

All buses with high voltage are set to be stable. The total compensation of SVG is the lowest. The decision scheme of the final test is shown in Table 6.

Fault 1 occurs on the highest level of the transmission bus between site *A* and site *B*. One of the decision schemes obtained by the model is action 19, i.e., five SVGs compensate 20 Mvar, 50 Mvar, 70 Mvar, 50 Mvar and 70 Mvar, respectively. It can be seen that the SVGs are distributed. The compensation value of each SVG is smaller than that of only two SVGs. The desired distributed setting of the reactive power compensation device is achieved. The stability result is shown in Figure 11. The time of returning to a stable state is about 550 cycles.

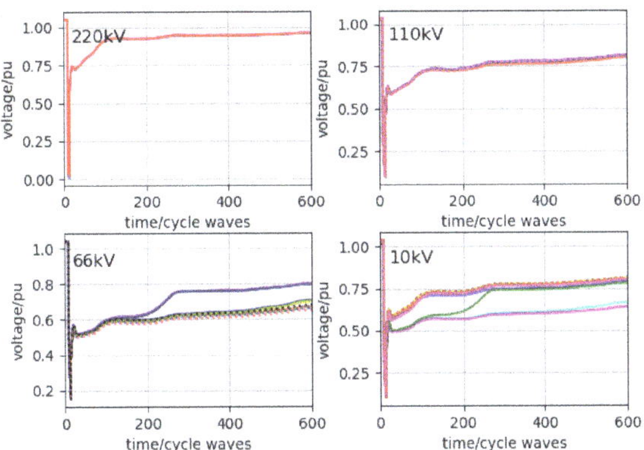

Figure 11. Fault 1 Action 19: Bus positive sequence voltages after SVG compensates 20 Mvar, 50 Mvar, 70 Mvar, 50 Mvar and 70 Mvar, respectively.

Fault 2 occurs on the high-voltage bus with 220 kV at site *B*. One of the decision schemes given by the model is action 34, that is, five SVGs compensate 20 Mvar, 50 Mvar, 80 Mvar, 50 Mvar and 60 Mvar, respectively. The stability result is shown in Figure 12. The time of returning to a state of stability is about 250 cycles.

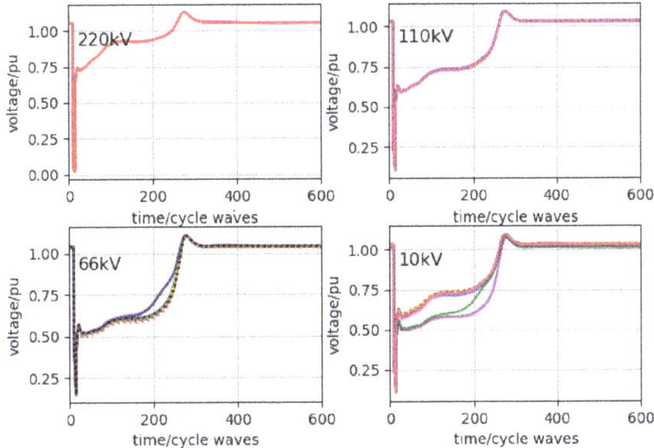

Figure 12. Fault 2 Action 34: Bus positive sequence voltages after SVG compensates 20 Mvar, 50 Mvar, 80 Mvar, 50 Mvar and 60 Mvar, respectively.

Fault 3 occurs on the high-voltage bus with 220 kV at site *B*. One of the decision schemes given by the model is action 389, i.e., five SVGs compensate 40 Mvar, 50 Mvar, 60 Mvar, 60 Mvar and 50 Mvar, respectively. The time of returning to a stable state is about 295 cycles.

Fault 4 occurs on the high-voltage bus with 220 kV at site *B*. One of the decision schemes given by the model is action 34, i.e., five SVGs compensate 20 Mvar, 50 Mvar, 80 Mvar, 50 Mvar and 60 Mvar, respectively. The time of returning to the state of stability is about 50 cycles.

From the above analysis, the difference between the high voltage bus with 220 kV and 110 kV in the initial stage of training is not distinguished by the model. The scheme with the same total compensation amount is selected. Although the requirement of restoring stability can be met, the whole EI is subject to the impact of smaller short-circuit faulting with lower bus voltage level. In the later stage of training, a scheme for fault 4 to fault 8 is given by the model. Five SVGs compensate 30 Mvar, 60 Mvar, 60 Mvar, 50 Mvar and 50 Mvar, respectively. At this point, the total compensation amount is 250 Mvar. Fault 4 is set, and then action 241 is executed. The stability result is shown in Figure 13. The time of returning to a stable state is about 60 cycles. A better optimal decision scheme can be given by the model though training.

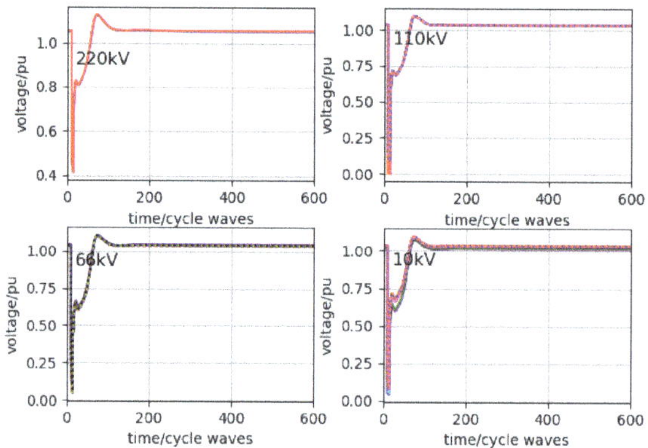

Figure 13. Fault 4 Action 241: Bus positive sequence voltages after SVG compensates 30 Mvar, 60 Mvar, 60 Mvar, 50 Mvar and 50 Mvar, respectively.

However, in the previous setting of the reward value calculation formula, only the requirement that buses with high voltage grade finally restore stability is considered. The time of returning to a stable state is not considered. Thereby, we can see from Figure 11 that although the strategy given by the algorithm ultimately achieves system stability, it is time consuming. The stable operation of the grid and the usage of the customer side's load would also be affected. The calculation of rewards is changed by the choice. The time of returning to the stable state is taken into account. New and different strategies are given by the algorithm in the final experimental results. One of these strategies is action 673, i.e., five SVGs compensate 50 Mvar, 70 Mvar, 60 Mvar, 50 Mvar and 50 Mvar, respectively. Although the total compensation is slightly higher than the previous strategy, it can be seen that the compensation of each SVG is distributed more evenly.

Fault 1 is set, and action 673 is executed. The stable result is shown in Figure 14. When five SVGs compensate 50 Mvar, 70 Mvar, 60 Mvar, 50 Mvar and 50 Mvar, respectively, the time of returning to a stable state is about 200 cycles.

Based on the comparison of the above experimental results, it can be seen that the strategy proposed by the final algorithm meets the requirement of bus voltage stability. Meanwhile, the SVG presents distributed settings, and the output compensation is distributed uniformly. In addition, the distributed SVG conducts reactive compensation with a shorter timeframe, which greatly improves the efficiency of decision-making compared with conventional manual operations. The distributed SVG obtains stability within 200 cycles, which meets the requirement for secure and stable operation of the EI.

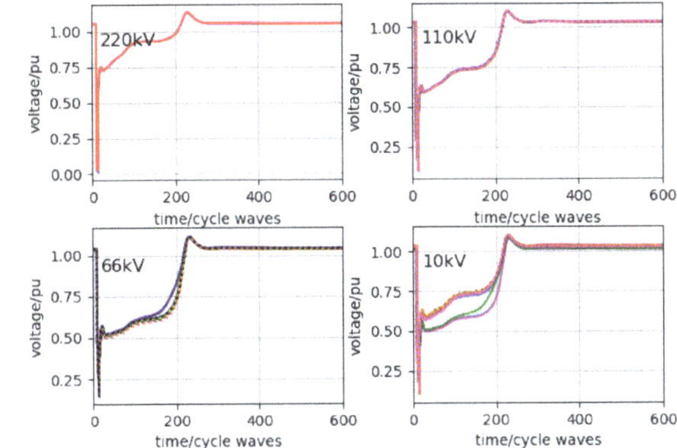

Figure 14. Fault 1 Action 673: Bus positive sequence voltages after SVG compensates 50 Mvar, 70 Mvar, 60 Mvar, 50 Mvar and 50 Mvar, respectively.

5. Conclusions

Based on EI architecture, this study proposes a decision optimization algorithm based on state judgment, in order to realize efficient, safe and stable operation of the EI. The experimental results show that the deep CNN is superior to the conventional machine learning algorithms with regards to feature extraction and prediction accuracy. In addition, a data batch processing toolkit based on BPA is developed to realize semi-automatic data batch processing, which improves the efficiency of data processing. Based on the stable state judgment, a deep reinforcement learning algorithm is proposed to optimize the reactive power compensation decision of EI. The experimental results not only show that this algorithm can achieve the system stability target, but can also fulfils the expectation of distributed reactive compensation and minimization of total reactive compensation.

Currently, a large number of simulation data can be generated off-line for training. The simulation and state feedback of continuous action changes cannot be realized. The action has to be discretized. In the future, it is necessary to realize the direct interface between simulation software and the deep learning platform, such that real-time simulation can be performed, and deep reinforcement learning algorithm can be used for continuous action learning and training.

Author Contributions: The work presented here was carried out through the cooperation of all authors. W.Z. and J.C. conceived the scope of the paper; W.Z. conceived the analysis and performed the simulations; H.H. and Z.X. wrote the paper; J.C. acquired the funding and performed revisions before submission. All authors read and approved the manuscript.

Funding: This work was supported in part by National Natural Science Foundation of China (grant No. 61472200) and Beijing Municipal Science & Technology Commission (grant No. Z161100000416004).

Conflicts of Interest: The authors declare no conflict of interest.

References

1. Liu, J.; He, D.; Wei, Q.; Yan, S. Energy storage coordination in energy internet based on multi-agent particle swarm optimization. *Appl. Sci.* **2018**, *8*, 1520. [CrossRef]
2. Cao, J.; Hua, H.; Ren, G. Energy use and the internet. In *The SAGE Encyclopedia of the Internet*; Sage: Newbury Park, CA, USA, 2018; pp. 344–350.
3. Wang, K.; Yu, J.; Yu, Y.; Qian, Y. A survey on energy internet: Architecture, approach, and emerging technologies. *IEEE Syst. J.* **2017**, *12*, 1–14. [CrossRef]
4. Hua, H.; Hao, C.; Qin, Y.; Cao, J. A class of control strategies for energy internet considering system robustness and operation cost optimization. *Energies* **2018**, *11*, 1593. [CrossRef]

5. Qiao, H.; Tian, J.; Tian, Z.; Qi, W.; Liu, C.; Li, X.; Zhu, H. An information security risk assessment algorithm based on risk propagation in energy internet. In Proceedings of the 2017 IEEE Conference on Energy Internet and Energy System Integration (EI2), Beijing, China, 26–28 November 2017; pp. 1–6.

6. Hua, H.; Qin, Y.; Hao, C.; Cao, J. Stochastic optimal control for energy Internet: A bottom-up energy management approach. *IEEE Trans. Ind. Inform.* **2019**, *15*, 1788–1797. [CrossRef]

7. Hua, H.; Qin, Y.; Geng, J.; Hao, C.; Cao, J. Robust mixed H_2/H_∞ controller design for energy routers in energy Internet. *Energies* **2019**, *12*, 340. [CrossRef]

8. Cao, Y.; Li, Q.; Tan, Y.; Li, Y.; Chen, Y.; Shao, X.; Zou, Y. A comprehensive review of energy internet: Basic concept, operation and planning methods, and research prospects. *J. Mod. Power Syst. Clean Energy* **2018**, *6*, 1–13. [CrossRef]

9. Yang, G.; Cao, J.; Hua, H.; Zhou, Z. Deep learning-based distributed optimal control for wide area energy Internet. In Proceedings of the 2nd IEEE International Conference on Energy Internet, Beijing, China, 20–22 October 2018; pp. 292–297.

10. Shang, Y. Resilient multiscale coordination control against adversarial nodes. *Energies* **2018**, *11*, 1844. [CrossRef]

11. Qiu, Y.; Wu, H.; Song, Y.; Wang, J. Global approximation of static voltage stability region boundaries considering generator reactive power limits. *IEEE Trans. Power Syst.* **2018**, *33*, 5682–5691. [CrossRef]

12. Hua, H.; Cao, J.; Yang, G.; Ren, G. Voltage control for uncertain stochastic nonlinear system with application to energy internet: Non-fragile robust H_∞ approach. *J. Math. Anal. Appl.* **2018**, *463*, 93–110. [CrossRef]

13. Song, S.; Yoon, M.; Jang, G. Analysis of six active power control strategies of interconnected grids with VSC-HVDC. *Appl. Sci.* **2019**, *9*, 183. [CrossRef]

14. Wang, L.; Kerrouche, K.D.E.; Mezouar, A.; Van Den Bossche, A.; Draou, A.; Boumediene, L. Feasibility study of wind farm grid-connected project in Algeria under grid fault conditions using d-facts devices. *Appl. Sci.* **2018**, *8*, 2250. [CrossRef]

15. Lu, J.Z.; Wu, C.P.; Tan, Y.J.; Zhu, S.G.; Sun, Y.C. Research of large-capacity low-cost DC Deicer with reactive power compensation. *IEEE Trans. Power Deliv.* **2018**, *33*, 3036–3044. [CrossRef]

16. Zhang, S.; Zhang, D.; Zhang, Y.; Cao, J. The research on smart power consumption technology based on big data. In Proceedings of the International Conference on Smart Grid and Clean Energy Technologies, Chengdu, China, 19–22 October 2016; pp. 12–18.

17. Lojda, J.; Podivinsky, J.; Kotasek, Z.; Krcma, M. Data types and operations modifications: A practical approach to fault tolerance in HLS. In Proceedings of the East-West Design & Test Symposium, Novi Sad, Serbia, 29 Septemper–2 October 2017; pp. 1–6.

18. Kundur, P. *Power System Stability and Control*; McGraw Hill Education: New York, NY, USA, 2002.

19. Chiang, H. *Direct Methods for Stability Analysis of Electric Power Systems: Theoretical Foundation, BCU Methodologies, and Applications*; John Wiley & Sons: Hoboken, NJ, USA, 2010.

20. Zhu, L.; Lu, C.; Dong, Z.Y.; Hong, C. Imbalance learning machine based power system short-term voltage stability assessment. *IEEE Trans. Ind. Inform.* **2017**, *13*, 2533–2543. [CrossRef]

21. Gomez, F.R.; Rajapakse, A.D.; Annakkage, U.D.; Fernando, I.T. Support vector machine-based algorithm for post-fault transient stability status prediction using synchronized measurements. *IEEE Trans. Power Syst.* **2011**, *26*, 1474–1483. [CrossRef]

22. Duraipandy, P.; Devaraj, D. Extreme learning machine approach for on-line voltage stability assessment. In *Swarm, Evolutionary, and Memetic Computing*; Springer International Publishing: Berlin, Germany, 2013; Volume 8298, pp. 397–405.

23. Dash, P.K.; Barik, S.K.; Patnaik, R.K. Detection and classification of islanding and nonislanding events in distributed generation based on fuzzy decision tree. *J. Control Autom. Electr. Syst.* **2014**, *25*, 699–719. [CrossRef]

24. Bulac, C.; Triştiu, I.; Mandiş, A.; Toma, L. On-line power systems voltage stability monitoring using artificial neural networks. In Proceedings of the International Symposium on Advanced Topics in Electrical Engineering, Bucharest, Romania, 7–9 May 2015; pp. 622–625.

25. Zhang, R.; Xu, Y.; Dong, Z.Y.; Zhang, P.; Wong, K.P. Voltage stability margin prediction by ensemble based extreme learning machine. In Proceedings of the 2013 IEEE Power & Energy Society General Meeting, Vancouver, BC, Canada, 21–25 July 2013.

26. Zhao, W.; Du, S. Spectral–spatial feature extraction for hyperspectral image classification: A dimension reduction and deep learning approach. *IEEE Trans. Geosci. Remote Sens.* **2016**, *54*, 4544–4554. [CrossRef]

27. Hinton, G.E. Deep belief networks. *Scholarpedia* **2009**, *4*, 5947. [CrossRef]

28. Wen, L.; Gao, L.; Li, X.A. New deep transfer learning based on sparse auto-encoder for fault diagnosis. *IEEE Trans. Syst. Man Cybern. Syst.* **2017**, *49*, 136–144. [CrossRef]

29. Mikolov, T.; Karafiát, M.; Burget, L.; Cernocky, J.; Khudanpur, S. Recurrent neural network based language model. In Proceedings of the Conference of the International Speech Communication Association, DBLP, Makuhari, Chiba, Japan, 26–30 September 2010; pp. 1045–1048.

30. Vincent, P.; Larochelle, H.; Lajoie, I.; Bengio, Y.; Manzagol, P. Stacked denoising autoencoders: Learning useful representations in a deep network with a local denoising criterion. *J. Mach. Learn. Res.* **2010**, *11*, 3371–3408.

31. Kim, P. Convolutional neural network. In *MATLAB Deep Learning*; Apress: New York, NY, USA, 2017.

32. Wang, D.L.; Sun, Q.Y.; Li, Y.Y.; Liu, X.R. Optimal energy routing design in energy internet with multiple energy routing centers using artificial neural network-based reinforcement learning method. *Appl. Sci.* **2019**, *9*, 520. [CrossRef]

33. Hua, H.; Qin, Y.; Hao, C.; Cao, J. Optimal energy management strategies for energy internet via deep reinforcement learning approach. *Appl. Energy* **2019**, *239*, 598–609. [CrossRef]

34. Ding, L.; Li, S.; Gao, H.; Chen, C.; Deng, Z. Adaptive partial reinforcement learning neural network-based tracking control for wheeled mobile robotic systems. *IEEE Trans. Syst. Man Cybern. Syst.* **2018**, *99*, 1–12. [CrossRef]

35. Wu, J.; He, H.; Peng, J.; Li, Y.; Li, Z. Continuous reinforcement learning of energy management with deep Q network for a power split hybrid electric bus. *Appl. Energy* **2018**, *222*, 799–811. [CrossRef]

36. Huang, Q.; Vittal, V. Application of electromagnetic transient-transient stability hybrid simulation to fidvr study. *IEEE Trans. Power Syst.* **2016**, *31*, 2634–2646. [CrossRef]

37. Li, Y.; Yuan, S.; Liu, W.; Zhang, G. A fast method for reliability evaluation of ultra high voltage AC/DC system based on hybrid simulation. *IEEE Access* **2018**, *6*, 19151–19160. [CrossRef]

38. Hu, B.; Dixon, P.C.; Jacobs, J.V.; Dennerlein, J.T.; Schiffman, J.M. Machine learning algorithms based on signals from a single wearable inertial sensor can detect surface- and age-related differences in walking. *J. Biomech.* **2018**, *71*, 37–42. [CrossRef]

39. Shang, Y. False positive and false negative effects on network attacks. *J. Stat. Phys.* **2018**, *170*, 141–164. [CrossRef]

Article

Short-Term Electric Power Demand Forecasting Using NSGA II-ANFIS Model

Aydin Jadidi *, Raimundo Menezes, Nilmar de Souza and Antonio Cezar de Castro Lima

Department of Electrical Engineering, Polytechnic School, Federal University of Bahia,
Salvador 40210-630, Brazil; raimundo.menezes@gmail.com (R.M.); nilmarufrb@gmail.com (N.d.S.);
acdcl@ufba.br (A.C.d.C.L.)
* Correspondence: aydin.jadidi@gmail.com; Tel.: +55-71-3283-9477

Received: 13 January 2019; Accepted: 8 April 2019; Published: 17 May 2019

Abstract: Load forecasting is of crucial importance for smart grids and the electricity market in terms of the meeting the demand for and distribution of electrical energy. This research proposes a hybrid algorithm for improving the forecasting accuracy where a non-dominated sorting genetic algorithm II (NSGA II) is employed for selecting the input vector, where its fitness function is a multi-layer perceptron neural network (MLPNN). Thus, the output of the NSGA II is the output of the best-trained MLPNN which has the best combination of inputs. The result of NSGA II is fed to the Adaptive Neuro-Fuzzy Inference System (ANFIS) as its input and the results demonstrate an improved forecasting accuracy of the MLPNN-ANFIS compared to the MLPNN and ANFIS models. In addition, genetic algorithm (GA), particle swarm optimization (PSO), ant colony optimization (ACO), differential evolution (DE), and imperialistic competitive algorithm (ICA) are used for optimized design of the ANFIS. Electricity demand data for Bonneville, Oregon are used to test the model and among the different tested models, NSGA II-ANFIS-GA provides better accuracy. Obtained values of error indicators for one-hour-ahead demand forecasting are 107.2644, 1.5063, 65.4250, 1.0570, and 0.9940 for RMSE, RMSE%, MAE, MAPE, and R, respectively.

Keywords: electric load forecasting; non-dominated sorting genetic algorithm II; multi-layer perceptron; adaptive neuro-fuzzy inference system; meta-heuristic algorithms

---

1. Introduction

Planning electricity systems in terms of generation, transmission, and distribution relies on generation and load forecasting. In addition to economic load dispatching, unit commitment, and price forecasting which, are interests of the electricity market, electrical load forecasting is important in terms of risk reduction for the power grid. In addition, unbalanced supply/demand caused by inaccurate forecasts in traditional electricity generation systems [1] has led to integration of advanced communication technologies into traditional grids, which are known as smart girds (Figure 1). Smart grids engage the customer in the decision-making process and, in a larger view, decisions are made based on the flow and exchange of information [2]. However, there are challenges to ensure that smart grids are economically beneficial, such as closing the gap between demand and supply, and fuel resource planning. All these factors highlight the importance of accurate electrical energy demand forecasts.

Diverse techniques have been applied in demand forecasting problems such as techniques based on time series and regression analysis [3–5]. However, because of the non-linear nature of the problem, techniques based on artificial neural networks and Adaptive Neuro-Fuzzy Inference System (ANFIS) are more popular [6–11]. As an example, Barak and Sadegh [12] proposed a hybrid ARIMA-ANFIS model for forecasting of the annual energy consumption of Iran. ARIMA outputs were used to forecast the energy consumption, using different ANFIS structures. According to the results, the ARIMA-ANFIS model gave more accurate forecasts compared to the ARIMA and ANFIS models. As the final step, meta-heuristic algorithms were employed to increase the accuracy of the ANFIS. The research does not develop a strategy for large data sets and input selection.

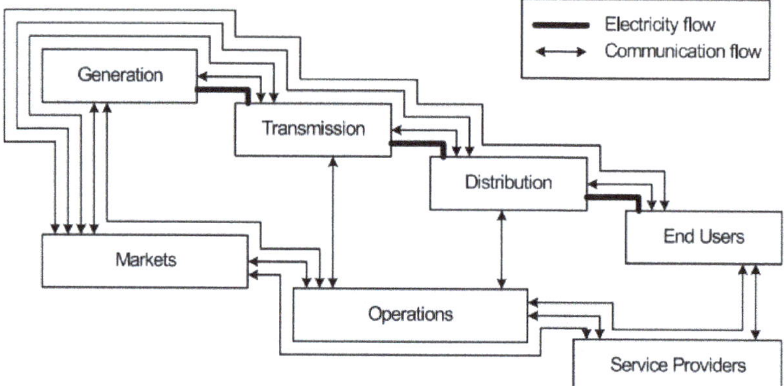

Figure 1. Conceptual diagram of smart grid [2].

In another piece of research conducted by Hooshmand et al. [13], a wavelet transform (WT) and an artificial neural network (ANN) were used for primary load forecasting where the inputs are meteorological parameters and previous values of the electric load. The ANFIS was employed to improve the forecasting results. However, the research does not introduce an approach for input selection and the capability of the evolutionary algorithms for optimizing the ANFIS has not been investigated. In another similar model, Panapakinis and Dagoumas [14] proposed a wavelet transform-ANFIS-GA-neural network model for natural gas demand forecasting. The original signal was decomposed by WT and used as ANFIS inputs. After optimizing ANFIS parameters with GA, output of ANFIS was fed into the neural network. The model does not seem to be efficient in case of multiple inputs since feature selection approach has not been developed.

A difference seasonal auto-regressive integrated moving average (diff-SARIMA), neural network, ANFIS, and DE combined method was used by Yang et al. [15] for short-term electricity demand forecasting of New South Wales in Australia. The proposed combined model presented better results than SARIMA, neural network, and ANFIS models. Parameters of the ANFIS were optimized using the DE method. Identical to the articles mentioned earlier, the research does not present a strategy for input selection. Moon et al. [16] proposed a hybrid of random forest and multi-layer perceptron for daily energy demand forecasting of a university campus. A decision tree was employed to classify the data into date, day of the week, holiday, and academic year. Furthermore, an approach was developed for considering the effect of the temperature in energy consumption and classifying the days of the week. However, the algorithm might not easily adapt to availability of the other parameters to be considered.

All the issues mentioned earlier motivated the current study to develop a forecasting strategy which: (i) can perform with any given dataset; (ii) is totally automatized; and (iii) provides a better accuracy.

Contribution

The current study aims to address solutions for data pre-processing and the input selection problem. As mentioned above, previous studies did not develop a robust model that is compatible with different datasets. The best combination of the input variables must be achieved before applying any data pre-processing or feature extraction techniques. The paper proposes a robust model which is capable of forecasting hourly electrical load demand with any given inputs. The inputs may include different combinations of previous demand values and weather parameters.

In addition, despite a combination of ANNs and ANFIS being discussed in previous research, training ANFIS was realized using a hybrid method. The proposed methodology employs meta-heuristic algorithms for ANFIS training and combines MLPNN, ANFIS, and meta-heuristics to increase the forecasting accuracy. Some research related to training ANFIS using meta-heuristics is given in [17–20].

The proposed methodology is described in Section 2 and the results are presented and discussed in Section 3. Finally, Section 4 is the conclusion of the present study.

2. Methodology

2.1. Data

Hourly electrical power demand data for Bonneville, Oregon, USA, from 2 July 2015 to 19 September 2017 provided by the US Energy Information Administration [21] were used to test and validate the model. The original data set was a $1 \times n$ matrix where n is the number of samples. After removing the matrix fields without reported values, a figure of the initial dataset was generated and presented in Figure 2.

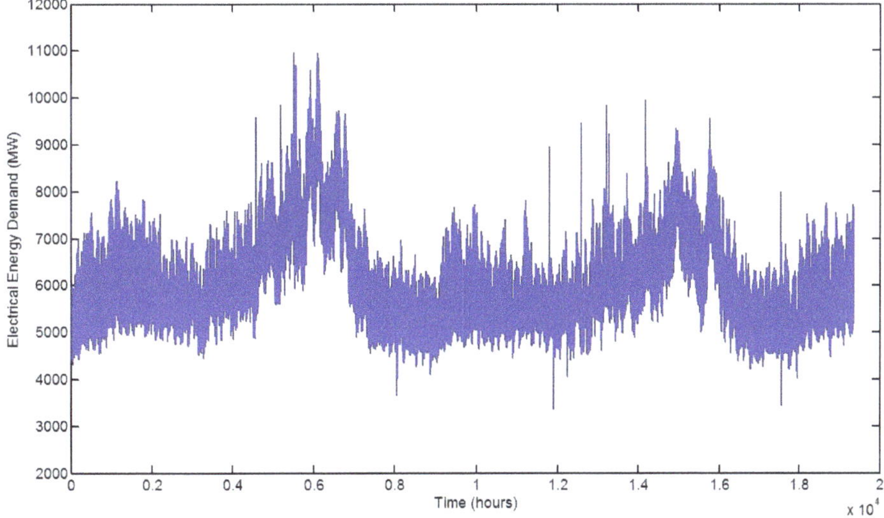

Figure 2. Electrical power demand data for Bonneville, Oregon, USA, from 2 July 2015 to 19 September 2017.

A new data set ($13 \times (n - 30)$ matrix) was created where 30 is the biggest considered delay. 13 is related to the applied probable delays of 1, 2, 3, 4, 5, 6, 24, 25, 26, 27, 28, 29, and 30 for creating a dataset with 13 variables. To obtain the best combination of the variables, a new dataset was fed into the NSGAII as described in Section 2.2. Outlier detection techniques can be applied for detection and removal of outliers. An example of these techniques is given in [22].

2.2. Proposed Algorithm

The created input dataset contains $(x - 1)$, $(x - 2)$, $(x - 3)$, $(x - 4)$, $(x - 5)$, $(x - 6)$, $(x - 24)$, $(x - 25)$, $(x - 26)$, $(x - 27)$, $(x - 28)$, $(x - 29)$, $(x - 30)$ variables where x is the actual electrical power demand. These are the inputs of the NSGA II which can generate results with any given input dataset and find the best combination of the inputs. Thus, other delays of the electrical power demand and weather parameters (if available) can be added to the inputs.

The proposed algorithm is a two-step forecasting process. In the primary forecasting step, a combination of the NSGAII and MLPNN was employed. MLPNN is the fitness function of the NSGA II to determine fitness of the input combinations in each iteration of the NSGAII. The output of the NSGAII was set to be the MLPNN with the best fitness. Therefore, the obtained MLPNN contains the best combination of the input variables among tested combinations in iterations of the NSGAII and is also the best-trained neural network.

As the second step, the obtained forecasted value from the first step was fed to the ANFIS. The result of this step is the final forecasted value of the electrical energy demand. Training of the ANFIS was realized using different algorithms, namely hybrid algorithm (combination of the backpropagation and least-square error), ACO, DE, GA, ICA, and PSO. Among applied algorithms for ANFIS training, GA demonstrated better performance in terms of the lower values of error indicators. The overall proposed approach is presented in Figure 3.

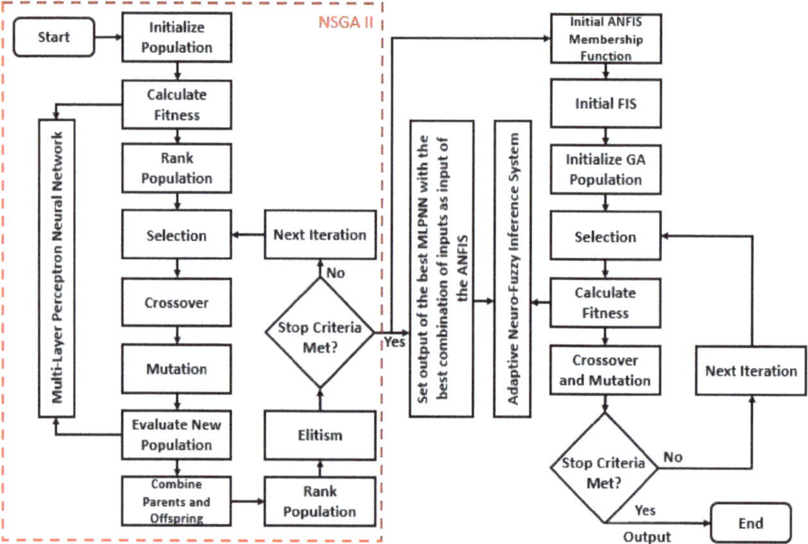

Figure 3. Proposed algorithm.

2.2.1. NSGA II

NSGA II [23] is an elitist multi-objective optimization algorithm that initializes by a random population and assigns a fitness value for each member of the population. After generating the offspring using crossover and mutation operators, a binary selection operator, based on fitness and crowding distance, is applied to the parent and offspring population for elitist selection. Crowding distance defines the distance of an individual to its neighbors and large crowding distance results in higher diversity, calculating the crowding distance begins by assigning distance to zero for each individual. Next, individuals are sorted based on objective function. After assigning infinite value to the boundaries ($I(d_1) = \infty, I(d_n) = \infty$), crowding distance of the mth objective function of the kth individual in front F_i for $k = 2$ to $(n-1)$ is calculated by Equation (1).

$$I(d_k) = I(d_k) = \frac{I(k+1) \cdot m - I(k-1) \cdot m}{f_m^{max} - f_m^{min}} \tag{1}$$

The algorithm preserves the best individuals from parent and offspring and continues until the stopping criteria is met. The elitism process is shown in Figure 4.

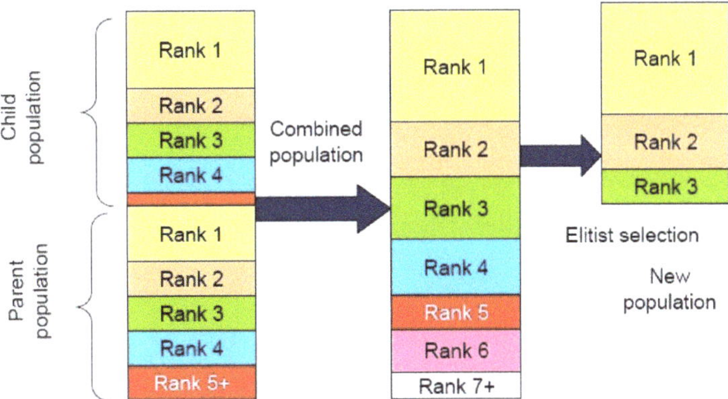

Figure 4. Elitism process of NSGA II.

2.2.2. MLPNN

Multi-layer perceptron neural network is a powerful tool for solving non-linear problems. It consists of an input layer, one or more hidden layers, and an output layer. Each layer contains artificial neurons and the neurons between layers are connected with an adaptable weight. The output of each neuron in each layer is multiplied by the adaptable weight and after passing through a transfer function becomes the input to the next-level neurons. In this research, tuning the weights, which is called the learning process, is realized by a backpropagation algorithm, namely the Levenberg-Marquardt algorithm. The structure of an MLPNN is shown in Figure 5.

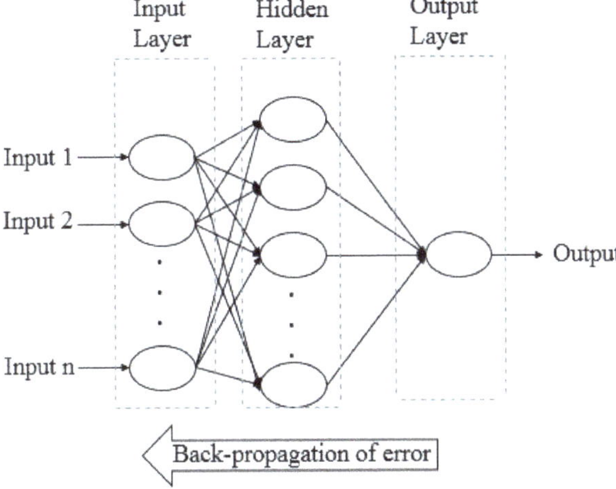

Figure 5. The structure of the MLPNN.

2.2.3. ANFIS

ANFIS was introduced by Jang [24] and is an artificial intelligence model that benefits from advantages of both fuzzy systems and ANNs. In this model, fuzzy inference systems (FIS) are determined by if-then rules and membership functions (MFs) where tuning the MFs are realized by ANNs. The three main types of FISs are the Takagi-Sugeno-Kang (TSK), Mamdani, and Tsumoto [25]. In this research, TSK FIS model was employed, which is more powerful at handling non-linear input-output relationships [26]. TSK uses the pattern of input and output data to create if-then rules. If-then rules for TSK model in a 2-input system is given in Equation (2).

$$If \quad x = A_1 \quad and \quad y = B_1 \Rightarrow f = p_1 x + q_1 x + r_1$$
$$If \quad x = A_2 \quad and \quad y = B_2 \Rightarrow f = p_2 x + q_2 x + r_2$$

(2)

where A_1, B_1 and A_2, B_2 are the MFs related to input x and input y respectively and $p_1, q_1, r_1, p_2, q_2, r_2$ are linear parameters of part-Then in TSK. Figure 6 shows the structure of a typical ANFIS model which contains 5 layers.

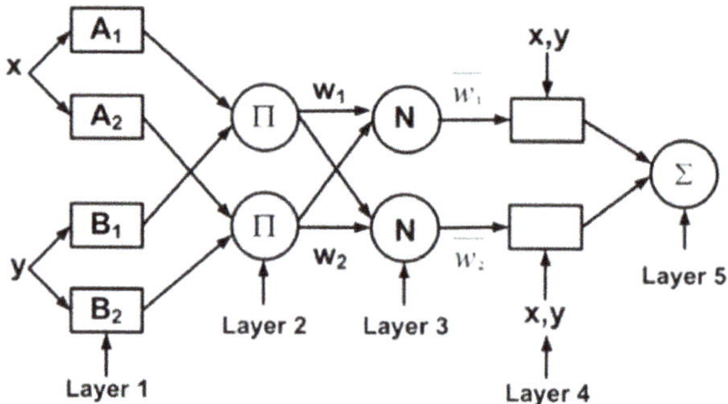

Figure 6. ANFIS architecture.

The fuzzification process is realized in the first layer. In this layer, nodes are square nodes with function presented in Equation (3).

$$O_i^1 = \mu A_i(x)$$

(3)

where i is the ith node in the layer, A_i is linguistic value of the node. As the membership function, a Gaussian membership function was employed (Equation (4)).

$$\mu A_i(x) = exp\left[-(\frac{x - c_i}{2a_i})^2\right]$$

(4)

where a_i and c_i are the parameters that define shape of the membership function. These parameters are tuned during the learning process. Layer 2 contains circle nodes that multiply the input signals and their output is the product of this multiplication which stands for firing strength of each rule as given in Equation (5).

$$\omega_i = \mu A_i(x) \times \mu B_i(y), i = 1, 2.$$

(5)

In layer 3, node i calculates the ratio of ith rule firing strength in respect to sum of all rules firing strength which, is a normalization process. Similar to layer 2, nodes are fixed in layer 3. Next, adaptive nodes in layer 4 calculate values of the rule of the consequent part and finally, layer 5 sums all outputs of layer 4 and contains only one node [27]. Mathematical description of layer 3, layer 4 and layer 5 are given in Equations (6)–(8) respectively.

$$\bar{\omega}_i = \frac{\omega_i}{\omega_1 + \omega_2}, i = 1, 2.$$

(6)

$$O_i^4 = \bar{\omega}_i f_i = \bar{\omega}_i (p_i x + q_i y + r_i).$$

(7)

$$O_i^5 = \sum_i \bar{\omega}_i f_i = \frac{\sum_i \omega_i f_i}{\sum_i \omega_i}.$$

(8)

ANFIS is trained using input-output data pairs. As seen in the ANFIS architecture and equations describing each layer, parameters in layer 1 and layer 4 can be tuned. This process defines the shape of the MFs (tuning of a_i and c_i) and specifies the fuzzy rules (tuning of p_i, q_i and r_i). Tuning of these parameters is realized according to an error criterion. Backpropagation algorithm can be employed for training ANFIS; however, due to slow convergence rate of the backpropagation algorithm and its tendency to be trapped in local minima, it is used combined with the least-square estimator. This combination is called the hybrid method and because of reduction of the dimensional search space, provides a faster convergence rate [28].

2.2.4. Genetic Algorithm (GA)

A genetic algorithm is a meta-heuristic algorithm based on natural selection. A continuous GA was used for this research, which consists of mutation, crossover, and selection. To distribute the initial population in solution space, the initial population is generated randomly. The solutions in each iteration are evolved until the stopping criteria is met.

3. Results and Discussion

3.1. Primary Forecasting Step

As the primary forecasting step, the created data matrix with 13 variables was fed into the NSGAII algorithm. The employed fitness function is an MLPNN with one hidden layer whose hidden layer size and transfer functions are 7, tansig, and tansig, respectively. Different combinations of the 13 variables (input vectors) were generated, and their fitness values were evaluated by the NSGAII in each iteration. Non-dominated solutions are the outputs of this step. Each output is a MLPNN, which contains the best non-dominated input vector as its input. The Pareto front of the NSGAII related to the generated non-dominant solutions is shown in Figure 7.

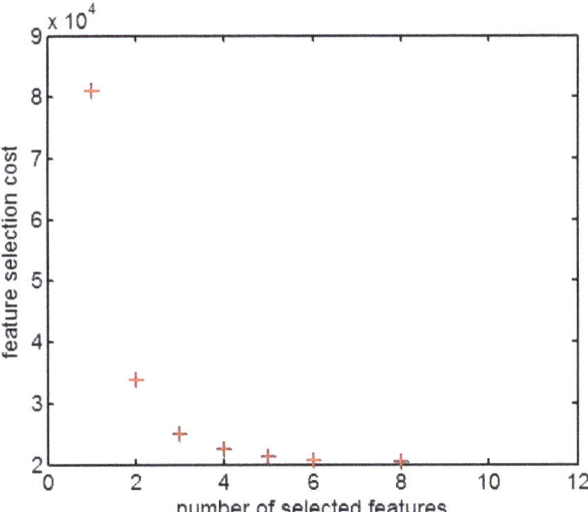

Figure 7. Pareto front of the NSGAII.

To demonstrate the importance of the feature selection, forecasting results of the best input vector generated by NSGAII was compared to the 13 variables 1 variable $(x - 1)$ input vectors, using MLPNN as the forecasting model. The results are presented in Table 1. Explanation of the employed error indicators can be found in Appendix A.

Table 1. Forecasting results for different input sets using MLPNN.

Input Set	RMSE (MW)	RMSE (%)	MAE (MW)	MAPE (%)	R
1 input	301.7724	3.8788	195.0414	3.1479	0.9523
Selected inputs by NSGA II	136.1048	1.7409	68.3688	1.0950	0.9904
All Inputs	160.5449	2.0636	69.4196	1.1029	0.9866

As seen in Table 1, using selected input vector generated by the NSGA II results in better forecasting accuracy. Furthermore, to compare the forecasting accuracy of the MLPNN and ANFIS models, the selected input vector was used as the input of the ANFIS models with different training algorithms. The results are given in Table 2.

Table 2. Forecasting results for each model.

Model	RMSE (MW)	RMSE (%)	MAE (MW)	MAPE (%)	R
ANFIS-ACOR	207.2688	2.8755	98.9818	1.5777	0.9775
ANFIS-Hybrid	194.6117	2.4017	86.8079	1.3838	0.9803
ANFIS-DE	202.3010	2.1925	100.7624	1.6181	0.9795
ANFIS-GA	190.1248	2.1810	97.4025	1.5438	0.9820
ANFIS-ICA	288.7683	3.8939	208.4114	3.3892	0.9577
ANFIS-PSO	195.7518	2.1215	83.5948	1.3360	0.9805

Comparing Table 1 to Table 2 it can be observed that using selected input vector, MLPNN model has a better forecasting accuracy compared to the ANFIS model. In addition, among tested ANFIS training algorithms, GA demonstrates better performance. Response surface of output versus input 1 and input 2 related to hybrid learning algorithm and GA learning algorithm (meta-heuristic with the best performance) are given in Figure 8 where input 1 and input 2 are $(x - 1)$ and $(x - 2)$ respectively.

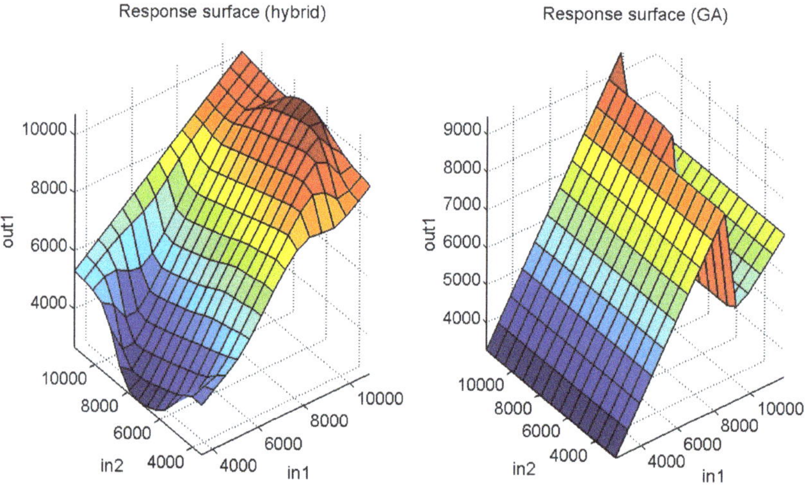

Figure 8. Response surface of output versus input 1 and input to for hybrid (**left**) and GA (**right**) learning algorithms.

3.2. Final Forecasting Step

In the primary forecasting step, variables were selected by the NSGAII and an output was generated which is an MLPNN with selected input vector. As the final step, output of the MLPNN was fed into the created ANFIS models with different training algorithms, to evaluate the ability of the ANFIS to increase the forecasting results of the MLPNN model. The results are given in Table 3.

Table 3. Forecasting results for each model.

Model	RMSE (MW)	RMSE (%)	MAE (MW)	MAPE (%)	R
MLPNN-ANFIS-ACOR	175.1901	1.8987	69.1994	1.1033	0.9842
MLPNN-ANFIS-Hybrid	142.6229	1.8243	66.5694	1.0603	0.9896
MLPNN-ANFIS-DE	158.2172	1.9138	68.2244	1.0967	0.9869
MLPNN-ANFIS-GA	107.2644	1.5063	65.4250	1.0570	0.9940
MLPNN-ANFIS-ICA	121.7895	1.5229	65.5095	1.0639	0.9922
MLPNN-ANFIS-PSO	148.1894	2.0559	66.8190	1.0641	0.9886

As seen in Table 3, the combination of the MLPNN and ANFIS models improves the forecasting accuracy and MLPNN-ANFIS models and demonstrates lower error rates compared to the MLPNN and ANFIS models. Furthermore, all error indicators of RMSE, MAE, MAPE, and R related to MLPNN-ANFIS-GA model are lower than MLPNN-ANFIS-GA model. Thus, GA has a better performance in ANFIS training compared to the hybrid method. Figure 9 presents the errors and absolute percentage error (APE) for ANFIS-GA and MLPNN-ANFIS-GA models to better demonstrate the increased forecasting accuracy using a combination of MLPNN and ANFIS.

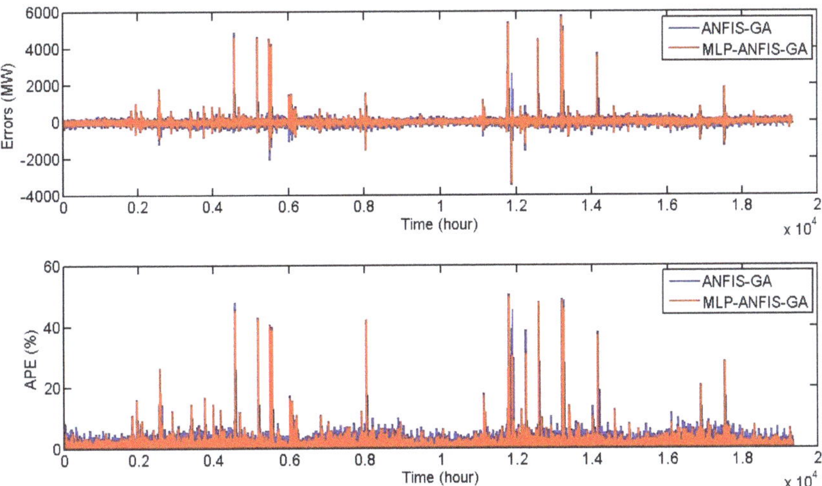

Figure 9. Errors and APE (%) for ANFIS-GA and MLP-ANFIS-GA models.

For the purposes of better illustrating the accuracy of the tested models, error indication of the correlation coefficient for MLPNN, ANFIS-Hybrid, ANFIS-GA, and MLPNN-ANFIS-GA models are presented in Figure 10 while error indicators for these models are presented in Table 4.

Table 4. Error indicators for MLPNN, ANFIS-Hybrid, ANFIS-GA, and MLPNN-ANFIS-GA models using the selected input vector by NSGAII.

Model	RMSE (MW)	RMSE (%)	MAE (MW)	MAPE (%)	R
MLPNN	136.1048	1.7409	68.3688	1.0950	0.9904
ANFIS-Hybrid	194.6117	2.4017	86.8079	1.3838	0.9803
ANFIS-GA	190.1248	2.1810	97.4025	1.5438	0.9820
MLPNN-ANFIS-GA	107.2644	1.5063	65.4250	1.0570	0.9940

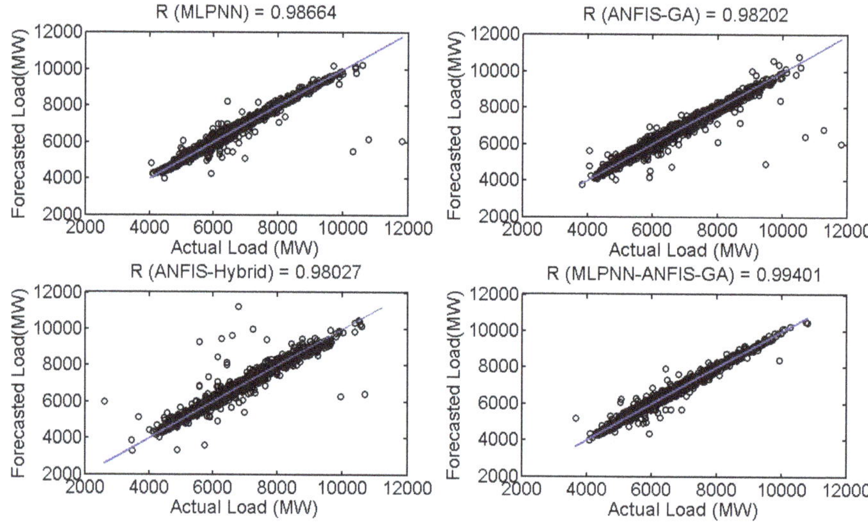

Figure 10. Correlation coefficient for MLPNN, ANFIS-Hybrid, ANFIS-GA, and MLPNN-ANFIS-GA models.

As seen, the MLPNN-ANFIS-GA model provides the best correlation coefficient and lower error rates in terms of the RMSE, MAE, and MAPE among the tested models. The targets (actual load) and the final forecasting results for a one-day region and a one-week region are presented in Figures 11 and 12 respectively.

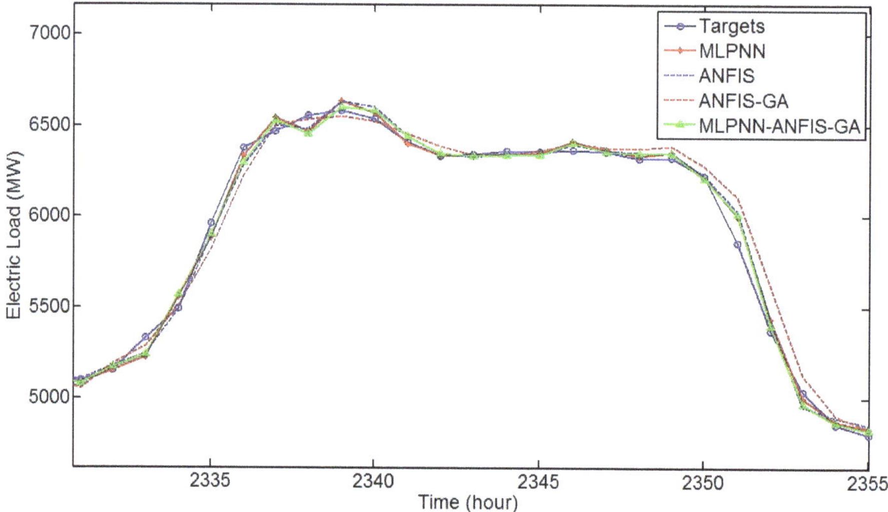

Figure 11. Forecasting results for a one-day region.

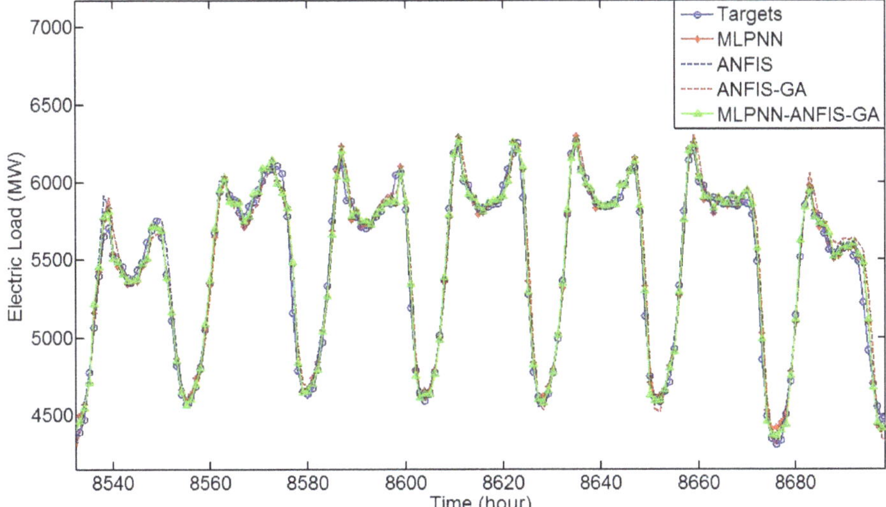

Figure 12. Forecasting results for a one-week region.

4. Conclusions

In all forecasting problems, input parameter selection is of great importance. The developed methodology in the current research proposes a solution to the trial-and-error approach employed by previous research in a demand forecasting field. The developed approach is compatible with any given dataset and can perform in cases of addition or removal of input variables. Assigning a multi-layer perceptron neural network as the output of NSGAII makes it possible to realize secondary processes automatically which, in the case of the current research, is using the ANFIS model to improve forecasting capability of the MLPNN.

Regarding processing time and computational complexity, since the most complex process is finding the input vector, once inputs are selected and the output of the NSGAII is generated, the algorithm can be used for online applications. It will be necessary to run the primary forecast part of the algorithm just in case of availability of new parameters, which can increase the forecasting accuracy.

According to the obtained results, while MLPNN has a better forecasting accuracy compared to the ANFIS, the combination of these two models reduces all forecasting error indicators. In addition, meta-heuristic algorithms are found to be suitable for training of the ANFIS. GA demonstrated better performance in terms of ANFIS training compared to the hybrid method. Among all tested models, the MLPNN-ANFIS-GA model presented lower error rates in terms of the RMSE, MAE, MAPE, and R.

Author Contributions: A.J. developed the methodology, analyzed data and generated results. R.M. and N.d.S. edited and re-wrote the manuscript drafts, and participated in generating results. A.C.d.C.L. supervised the research and approved the submitted manuscript.

Funding: This research has been funded by the Coordination for the Improvement of Higher Education Personnel (CAPES).

Conflicts of Interest: The authors declare no conflict of interest.

Abbreviations

The following abbreviations are used in this manuscript:

ACO	Ant Colony Optimization
ANFIS	Adaptive Neuro-Fuzzy Inference System
ANN	Artificial Neural Network
APE	Absolute Percentage Error

ARIMA	Auto-Regressive Integrated Moving Average
DE	Differential Evolution
FIS	Fuzzy Inference System
GA	Genetic Algorithm
ICA	Imperialistic Competitive Algorithm
MAE	Mean Absolute Error
MAPE	Mean Absolute Percentage Error
MF	Membership Function
MLPNN	Multi-Layer Perceptron Neural Network
NSGA II	Non-dominated Sorting Genetic Algorithm II
PSO	Particle Swarm Optimization
R	Correlation Coefficient
RMSE	Root Mean Square Error
TSK	Takagi-Sugeno-Kang
WT	Wavelet Transform

Appendix A

The indicators of absolute percentage error ($APE\%$), mean absolute error (MAE), mean absolute percentage error ($MAPE$), root mean squared error ($RMSE$), root mean squared error percentage ($RMSE\%$), and correlation coefficient (R) were used for evaluations of the model. The following equations describe these indicators:

$$APE = \frac{|y_i - x_i|}{x_i} \times 100 \tag{A1}$$

$$MAE = \frac{1}{n} \sum_{i=1}^{n} |x_i - y_i| \tag{A2}$$

$$MAPE = \frac{1}{N} \sum_{i=1}^{N} (\frac{|y_i - x_i|}{x_i} \times 100) \tag{A3}$$

$$RMSE = \sqrt{\frac{1}{n} \sum_{i=1}^{n} (x_i - y_i)^2} \tag{A4}$$

$$RMSE\% = (\frac{RMSE}{x_{max} - x_{min}}) \times 100 \tag{A5}$$

$$R = \frac{\sum_{i=1}^{n} (x_i - \bar{x})(y_i - \bar{y})}{\sqrt{\sum_{i=1}^{n} (x_i - \bar{x})^2 \sum_{i=1}^{n} (y_i - \bar{y})^2}} \tag{A6}$$

where x_i and $\bar{x}_i$ are the actual load value and mean of the actual load value and y_i and $\bar{y}_i$ are the forecasted load value and mean of the forecasted load value, respectively.

References

1. Hassan, S.; Khosravi, A.; Jaafar, J.; Khanesar, M.A. A systematic design of interval type-2 fuzzy logic system using extreme learning machine for electricity load demand forecasting. *Int. J. Electr. Power Energy Syst.* **2016**, *82*, 1–10. [CrossRef]
2. Ahmad, A.; Javaid, N.; Mateen, A.; Awais, M.; Khan, Z.A. Short-Term Load Forecasting in Smart Grids: An Intelligent Modular Approach. *Energies* **2019**, *12*, 164. [CrossRef]
3. Javed, F.; Arshad, N.; Wallin, F.; Vassileva, I.; Dahlquist, E. Forecasting for demand response in smart grids: An analysis on use of anthropologic and structural data and short term multiple loads forecasting. *Appl. Energy* **2012**, *96*, 150–160. [CrossRef]
4. Tascikaraoglu, A.; Sanandaji, B.M. Short-term residential electric load forecasting: A compressive spatio-temporal approach. *Energy Build.* **2016**, *111*, 380–392. [CrossRef]

5. Mori, H.; Kosemura, N. Optimal regression tree based rule discovery for short-term load forecasting. In Proceedings of the 2001 IEEE Power Engineering Society Winter Meeting, Conference Proceedings (Cat. No.01CH37194), Columbus, OH, USA, 28 January–1 February 2001; Volume 2; pp. 421–426. [CrossRef]

6. Hippert, H.S.; Pedreira, C.E.; Souza, R.C. Neural networks for short-term load forecasting: A review and evaluation. *IEEE Trans. Power Syst.* **2001**, *16*, 44–55. [CrossRef]

7. Rodrigues, F.; Cardeira, C.; Calado, J.M.F. The Daily and Hourly Energy Consumption and Load Forecasting Using Artificial Neural Network Method: A Case Study Using a Set of 93 Households in Portugal. *Energy Procedia* **2014**, *62*, 220–229. [CrossRef]

8. Beccali, M.; Cellura, M.; Brano, V.L.; Marvuglia, A. Short-term prediction of household electricity consumption: Assessing weather sensitivity in a Mediterranean area. *Renew. Sustain. Energy Rev.* **2008**, *12*, 2040–2065. [CrossRef]

9. Jadidi, A.; Menezes, R.; de Souza, N.; de Castro Lima, A.C. A hybrid GA–MLPNN Model for one-hour-ahead forecasting of the global horizontal irradiance in Elizabeth City, North Carolina. *Energies* **2018**, *11*, 2641. [CrossRef]

10. Saleh, A.E.; Moustafa, M.S.; Abo-Al-Ez, K.M.; Abdullah, A.A. A hybrid neuro-fuzzy power prediction system for wind energy generation. *Int. J. Electr. Power Energy Syst.* **2016**, *74*, 384–395. [CrossRef]

11. Nikolovski, S.; Baghaee, H.R.; Mlakic, D. ANFIS-Based Peak Power Shaving/Curtailment in Microgrids Including PV Units and BESSs. *Energies* **2018**, *11*, 2953. [CrossRef]

12. Barak, S.; Sadegh, S.S. Forecasting energy consumption using ensemble ARIMA–ANFIS hybrid algorithm. *Int. J. Electr. Power Energy Syst.* **2016**, *82*, 92–104. [CrossRef]

13. Hooshmand, R.A.; Amooshahi, H.; Parastegari, M. A hybrid intelligent algorithm based short-term load forecasting approach. *Int. J. Electr. Power Energy Syst.* **2013**, *45*, 313–324. [CrossRef]

14. Panapakidis, I.P.; Dagoumas, A.S. Day-ahead natural gas demand forecasting based on the combination of wavelet transform and ANFIS/genetic algorithm/neural network model. *Energy* **2017**, *118*, 231–245. [CrossRef]

15. Yang, Y.; Chen, Y.; Wang, Y.; Li, C.; Li, L. Modelling a combined method based on ANFIS and neural network improved by DE algorithm: A case study for short-term electricity demand forecasting. *Appl. Soft Comput.* **2016**, *49*, 663–675. [CrossRef]

16. Moon, J.; Kim, Y.; Son, M.; Hwang, E. Hybrid Short-Term Load Forecasting Scheme Using Random Forest and Multilayer Perceptron. *Energies* **2018**, *11*, 3283. [CrossRef]

17. Bakyani, A.E.; Sahebi, H.; Ghiasi, M.M.; Mirjordavi, N.; Esmaeilzadeh, F.; Lee, M.; Bahadori, A. Prediction of CO_2–oil molecular diffusion using adaptive neuro-fuzzy inference system and particle swarm optimization technique. *Fuel* **2016**, *181*, 178–187. [CrossRef]

18. Al-Dunainawi, Y.; Abbod, M.F.; Jizany, A. A new MIMO ANFIS-PSO based NARMA-L2 controller for nonlinear dynamic systems. *Eng. Appl. Artif. Intell.* **2017**, *62*, 265–275. [CrossRef]

19. Sohrabpoor, H. Analysis of laser powder deposition parameters: ANFIS modeling and ICA optimization. *Optik* **2016**, *127*, 4031–4038. [CrossRef]

20. Khosravi, A.; Nunes, R.; Assad, M.; Machado, L. Comparison of artificial intelligence methods in estimation of daily global solar radiation. *J. Clean. Prod.* **2018**, *194*, 342–358. [CrossRef]

21. U.S. Energy Information Adminstration (EIA) Home Page. Available online: https://www.eia.gov/ (accessed on 17 December 2017).

22. Ester, M.; Kriegel, H.P.; Sander, J.; Xu, X. A Density-Based Algorithm for Discovering Clusters in Large Spatial Databases with Noise. In Proceedings of the Second International Conference on Knowledge Discovery and Data Mining (KDD-96), Portland, OR, USA, 2–4 August 1996; pp. 226–231.

23. Deb, K.; Pratap, A.; Agarwal, S.; Meyarivan, T. A fast and elitist multiobjective genetic algorithm: NSGA-II. *IEEE Trans. Evol. Comput.* **2002**, *6*, 182–197. [CrossRef]

24. Jang, J.R. ANFIS: Adaptive-network-based fuzzy inference system. *IEEE Trans. Syst. Man Cybern.* **1993**, *23*, 665–685. [CrossRef]

25. Yaseen, Z.M.; Ebtehaj, I.; Bonakdari, H.; Deo, R.C.; Mehr, A.D.; Mohtar, W.H.M.W.; Diop, L.; El-Shafie, A.; Singh, V.P. Novel approach for streamflow forecasting using a hybrid ANFIS-FFA model. *J. Hydrol.* **2017**, *554*, 263–276. [CrossRef]

26. Karkevandi-Talkhooncheh, A.; Hajirezaie, S.; Hemmati-Sarapardeh, A.; Husein, M.M.; Karan, K.; Sharifi, M. Application of adaptive neuro fuzzy interface system optimized with evolutionary algorithms for modeling CO_2-crude oil minimum miscibility pressure. *Fuel* **2017**, *205*, 34–45. [CrossRef]

27. Mlakić, D.; Baghaee, H.R.; Nikolovski, S. A Novel ANFIS-based Islanding Detection for Inverter–Interfaced Microgrids. *IEEE Trans. Smart Grid* **2018**. [CrossRef]

28. Nayak, P.; Sudheer, K.; Rangan, D.; Ramasastri, K. A neuro-fuzzy computing technique for modeling hydrological time series. *J. Hydrol.* **2004**, *291*, 52–66. [CrossRef]

Article

Fuzzy Neural Network Control of Thermostatically Controlled Loads for Demand-Side Frequency Regulation

Zhengwei Qu, Chenglin Xu, Kai Ma * and Zongxu Jiao

School of Electrical Engineering, Yanshan University, Qinhuangdao 066004, China
* Correspondence: kma@ysu.edu.cn; Tel.: +86-335-838-7556

Received: 4 June 2019; Accepted: 24 June 2019; Published: 26 June 2019

Abstract: In this paper, a fuzzy neural network controller for regulating demand-side thermostatically controlled loads (TCLs) is designed with the aim of stabilizing the frequency of the smart grid. Specifically, the balance between power supply and demand is achieved by tracking the automatic generation control (AGC) signal in an electric power system. The particle swarm optimization (PSO) and error back propagation (BP) algorithms are used to optimize the control parameters and consequently reduce the tracking errors. The fuzzy neural network can be applied to solve load control problems in power systems, since its self-learning and associative storage functions can deal with the highly nonlinear relationship between input and output. Simulation results show the advantage of the fuzzy neural network control scheme in terms of frequency regulation error and consumer comfort.

Keywords: automatic generation control; fuzzy neural network control; thermostatically controlled loads; back propagation algorithm; particle swarm optimization

---

1. Introduction

Smart grid is a modern grid infrastructure with high efficiency, reliability, and safety, which is based on renewable energy, automatic control, and modern communication technology [1,2]. An ancillary service is indispensable in an electric service, and plays a vital role in providing strong support for power transmission and power system operation. Ancillary service includes services related to frequency stability. The stability of grid frequency is closely related to the operation of power market and the equipment safety of the power generation side and power consumption side. Ancillary service mainly includes: Frequency adjustment, automatic generation control, spinning reserve, and peaking services.

The mismatch between the power supply side and the area power consumption can affect the frequency of the power system. Hence, the load frequency regulation is necessary in power systems, in order to maintain power balance under normal conditions [3]. Grid frequency can be used to evaluate power quality. The main way to adjust the frequency of the power system is to change the generated output power and manage the loads in demand side. Reasonable control of temperature control loads can provide adjustable buffer energy for power systems [4]. A great deal of potential electrically-thermal energy is stored in all sorts of heat buffers equipment, such as heating, air conditioning units, and fridges [5]. The advantages of aggregated thermostatically controlled loads (TCLs) to take part in power grid frequency regulation are as follows: Firstly, the TCLs which can store massive energy are of wide distributions; secondly, the control method of them is simple, fast, and real-time; thirdly, the aggregated TCLs can generate a continuous reaction without considering the discrete characteristics of the individual load control [6]. The management of TCLs is one of frequency adjustment strategies with extremely high feasibility to ensure frequency stability and improve power quality.

Recently, the bilinear partial differential equation (PDE) model was developed to provide effective control of aggregated TCLs [7]. Two control methods, based on the combination of estimator and controller, were utilized to control TCLs and effectively track demand curve [8]. A centralized and distributed algorithm based on the state space model of heating ventilation and air-conditioning (HVAC) was proposed to reduce power fluctuations and improve satisfaction of demand side [9]. A model predictive control was described and applied to demand-side response services in [10]. A distributed model predictive control (DMPC) method based on an optimized aggregation model was applied to provide frequency regulation services [11]. A Fokker–Planck diffusion model and a direct load control algorithm were developed in [12]. A model of aggregate homogeneous TCLs with uniform variation of temperature set-point was developed, and a linear quadratic regulator (LQR) has been designed in [13]. In [14], the authors proposed a novel causal method based on a parametric second-order model to forecast the energy conservation. In [15], a linear optimization model was built to provide frequency regulation services for power systems while also providing short-term demand response management. In [16], the authors proposed several switched control strategies for aggregate HVACs to provide demand-side frequency regulation. Although various aggregation characteristics of TCLs have been extensively studied, modeling precision still needs to be farther improved. As a matter of fact, it is hard to establish accurate mathematical models or physical models for aggregated TCLs because of various assumptions and computational complexity in the modeling process.

In recent years, artificial intelligence has developed rapidly with the characteristics of bionics and intelligence. This paper proposes a fuzzy neural network control method which adjusts TCLs based on the input and output data of TCLs instead of the aggregated TCL model. This method can reduce tracking errors and computational complexity, because it draws the advantages–logic reasoning capability of fuzzy control and self-learning capability of the neural network [17]. This study has the following contributions:

- The fuzzy neural network can model the characteristics of aggregated TCLs in the network weight after training, which can meet the response of the system input and output without modeling the load characteristics.
- The combination of particle swarm optimization (PSO) and back propagation (BP) algorithms, which optimize the initial value of weight and membership degree parameters, can avoid the local minimum and accelerate effectively to convergence speed.

The rest of this paper is organized as follows. Section 1 introduces the thermal dynamics of individual TCLs and the frequency regulation problem. Section 2 describes the system structure, algorithm, and optimization of the fuzzy neural network control in detail. The simulation results are shown in Section 3. Finally, the conclusions are summarized in Section 4.

2. Problem Formulation

2.1. Individual TCL Characteristic

The TCLs can consume electric energy and release thermal energy, which can usually be stored and transferred [18]. For example, the simple TCLs usually have two work states, i.e., "on" or "off", and each corresponds to one power value, i.e., P_{rate} or 0. When there is excess power in generation side, the TCL is changed to the "on" state, then the electricity consumption increases and the transformed heat energy will also increase. When the power generation fails to meet users' needs, the TCLs are changed to the "off" state and the power demand will be reduced in order to stabilize the power grid frequency. The characteristics are described as follows [19]:

$$\dot{T}(t) = \frac{1}{CR}(T_a - T(t) - s(t)RP),$$ (1)

where T and T_a represent the indoor and outdoor temperature, respectively. P, R, and C denote energy transfer rate, thermal resistance, and thermal capacitance, respectively. $s(t)$ represents the switching state of loads.

The operation characteristics of an individual TCL are shown in Figures 1 and 2. The set-point regulation method is adopted for regulating TCLs to achieve the purpose of peak shaving and load shifting. In the figure, u is the temperature set-point change. T_+, T_s, and T_- on the left of the figure denote the upper boundary of temperature, the temperature set-point, and the lower boundary of temperature, respectively. $\triangle$ is the width of the temperature deadband, which denotes the difference between the upper and lower limits of temperature. According to the physical characteristics of TCLs, T_+, T_s, T_-, and $\triangle$ have the following relationship:

$$\begin{cases} T_+ = T_s + \triangle/2, \\ T_- = T_s - \triangle/2, \\ \triangle = T_+ - T_-. \end{cases}$$ (2)

The rising edge represents that the room temperature has a rise caused by a natural heat conduction process when the TCL is off. The falling edge denotes the temperature drops caused by the cooling process of the TCL when the TCL is turned on. In order to keep the room temperature near the temperature set-point, the TCL will be changed from an off to an on state when the temperature reaches the upper limit and changed from on to off state when the temperature reaches the lower limit in Figure 1. We can observe that the upper and lower bounds of temperature vary with the temperature set-point, but the deadband $\triangle$ keeps constant in Figure 2. Hence, when the temperature set-point changes, the TCL's switch state can be indirectly changed, and the TCL's running time will be extended or shortened. Therefore, prolonging or shortening the running time of TCL will change the demand-side power consumption, thus it is effective to maintain the stability of the power grid frequency.

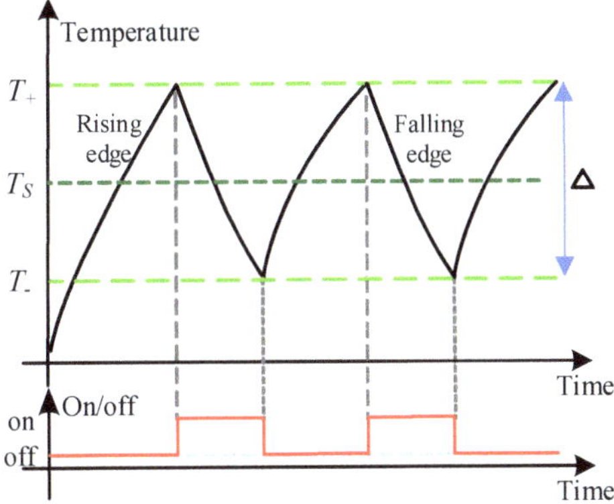

Figure 1. Operation characteristic of an individual thermostatically controlled load (TCL) ($u = 0$).

Suppose that N TCLs are used to provide the frequency regulation. When u is changed, the power consumption in demand side will be shifted, which contributes to the peak shaving and valley filling for the grid. The aggregated power consumption P_{total} could be expressed by

$$P_{total}(t) = \sum_{i=1}^{N} \frac{1}{\eta_i} s_i P_i,$$ (3)

where η_i (>1) is the efficiency coefficient of the ith TCL. s_i is a binary variable. The TCL is on when $s_i = 1$; and off when $s_i = 0$. P_i denotes the rated power of the ith load.

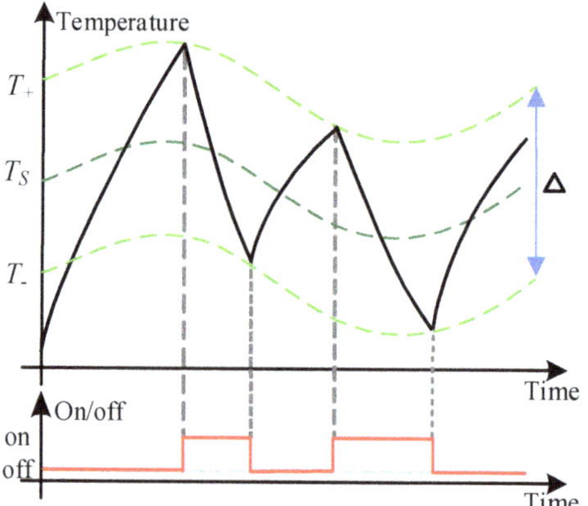

Figure 2. Operation characteristic of an individual TCL ($u \neq 0$).

2.2. Frequency Regulation Problem

In the power system, since the power is difficult to store in large quantities, the real-time power balance between the generation side and the demand side must be maintained which can achieve the goal of suppressing the grid frequency fluctuations, keeping the reliable power supply, avoiding accidents and hurting the user power equipment [20,21]. The aggregated TCLs can reliably provide the frequency regulation service by energy storage and flexible scheduling, which are mentioned by demand-side management. In that case, the power consumption of the TCLs tracks the frequency modulation power signal of the power grid by an effective controller and a control algorithm.

The frequency regulation signal P_{AGC}, which is a continuous change of positive and negative power signals and reflects the supply and demand deviation, thus, P_{AGC} needs to stacked on a baseline load signal P_{BL}—that will generate the actual tracking signal P_{target}, which can be tracked by the aggregated TCLs. Here, P_{BL} is selected as the rated power at one temperature set-point. It can be described as:

$$P_{BL} = \sum_{i=1}^{N} \frac{T_a - T_i^{set}}{\eta_i R_i},$$ (4)

$$P_{target} = P_{AGC} + P_{BL},$$ (5)

where N represents that the quantity of TCLs.

From Figure 3, we can observe that the inputs of the controller are the difference between the actual tracking signal P_{target} and the power consumption P_{total}, which are the tracking error signal

$e = P_{\text{target}} - P_{\text{total}}$ and its differential *de*. The controller output is the temperature set-point change *u*. In the control scheme, since the fuzzy neural network controller looks like a black box with brilliant self-learning and self-adaptive ability, we do not need to model the aggregated TCLs. The fuzzy neural network can be trained to encode the internal characteristics of the controlled object into the initial value of the connection weight. The connection of BP and PSO algorithm is also used to optimize the parameters of the controller and achieve the goal of reducing load tracking errors, improving tracking performance of the power grid frequency signal, and providing better frequency regulation service.

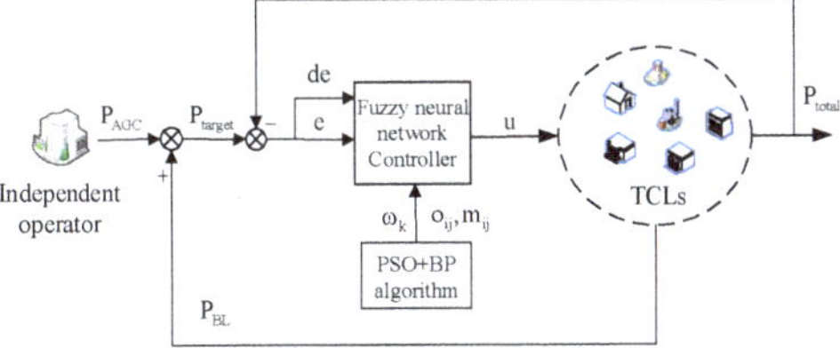

Figure 3. System structure diagram.

3. Fuzzy Neural Network Control Scheme

3.1. Fuzzy Neural Network Structure

The fuzzy neural network structure established in this paper has four layers, each layer is connected by weighted values. The multi-layer forward fuzzy neural network has the following advantages and characteristics for automatic control:

- The network can realize any complex nonlinear mapping. In this paper, there is a nonlinear relationship between the real-time tracking error and the temperature set-point, which is not easy to be formulated. Therefore, the network essentially implements a mapping function from input to output, which is especially suitable for solving complicated internal mechanism problems;
- The network has strong self-learning ability. First, it can learn the nonlinear relationship between the real-time tracking error and the temperature set-point—that is, the training process—then it can reflect the extracted "reasonable" solving rules to the connection coefficients between the layers. The appropriate connection coefficient has great influence on the control effect for frequency regulation;
- The network has certain promotion and generalization ability.

BP algorithm is adopted in the fuzzy neural network, its main feature is the forward transmission of the signal and the backward propagation of the error [22]. Since the transfer functions must satisfy the conditions of being differentiable everywhere, the hidden layer transfer function and the output layer activation function are Gaussian function and linear function, respectively. Fuzzy neural network mainly refers to the use of neural network structure to achieve fuzzy logic reasoning. Compared with a traditional neural network, the second layer can give specific physical meaning to tracking error, and the third layer is the fuzzy logic reasoning layer in the design of fuzzy neural network. The structure of the fuzzy neural network is shown in Figure 4 [23].

We define the error *e* and the error variation *de* as fuzzy linguistic variables, and these two fuzzy language variables contain five linguistic values, i.e., NB, NS, ZO, PS, and PB. Thus, there are 25 rules in the third layer of the fuzzy neural network structure. In the following, x_i^k represents the *i*th input of layer

k, and o_j^k represents the net input of the jth node of layer k, and y_j^k represents the output of the jth node of layer k. The input and output relationships of each layer node are as follows [24,25]:

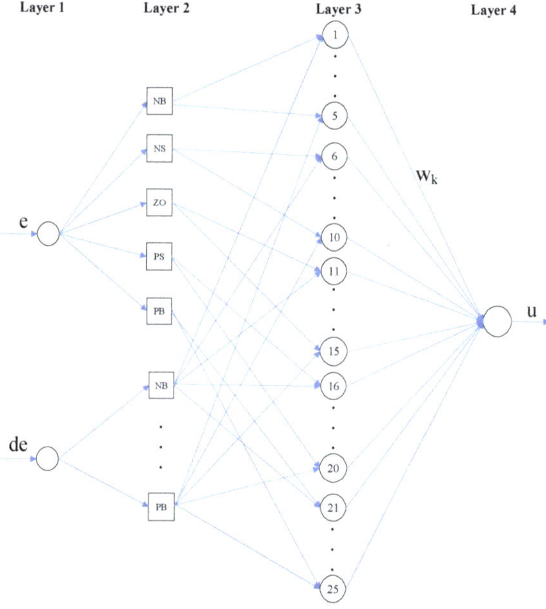

Figure 4. Fuzzy neural network structure diagram.

The fuzzy neural network has input layer, membership function layer, rule layer, and output layer from left to right in Figure 4. The input–output relationship in the input layer is

$$o_i^1 = s(i), \qquad i = 1, 2 \tag{6}$$

$$s = [e \quad de]', \tag{7}$$

$$y_i^1 = o_i^1. \qquad i = 1, 2 \tag{8}$$

The membership function layer is used to evaluate membership degree of each input component which belongs to the fuzzy set of each linguistic variable. The input–output relationship of this layer is

$$o_{ij}^2 = h(y_i^1), \qquad i = 1, 2 \quad j = 1, 2, \dots, 5, \tag{9}$$

$$y_k^2 = f(o_{ij}^2), \qquad k = 1, 2, \dots, 10, \tag{10}$$

where o_{ij}^2 represents the input of the jth fuzzy set of the ith fuzzy variable. For the same fuzzy concept, the membership functions can be different. Although the form is not exactly the same, the functions follow normal distribution which can reflect the fuzzy information that is processed by fuzzy concept when solving problems. For example, if the membership function of normal distribution is adopted, the result is shown as follows:

$$h(x) = -\frac{(x - m_j)^2}{\sigma_j^2}, \tag{11}$$

$$f(x) = exp(x),\tag{12}$$

where m_j and σ_j are parameters of Gauss membership function of x. Hence, the input and output relationships of the second layer nodes are shown as

$$o_{ij}^2 = h(y_i^1) = -\frac{(y_i^1 - m_{ij})^2}{\sigma_{ij}^2},\tag{13}$$

$$y_j^2 = f(o_{ij}^2) = exp(o_{ij}^2).\tag{14}$$

The third layer denotes fuzzy rule sets, every node denotes a fuzzy rule. This layer has 25 nodes. The input–output relationships of the third layer nodes are shown in Equations (15) and (16):

$$o_n^3 = y_i^2 * y_j^2, \qquad i = 1, 2, ..., 5 \quad j = 1, 2, ..., 5\tag{15}$$

$$y_n^3 = o_n^3. \qquad n = 1, 2, ..., 25.\tag{16}$$

The fourth layer denotes controller output, i.e., the optimal result obtained by fuzzy neural network:

$$o_1^4 = \frac{\sum\limits_{k=1}^{25} w_k * y_k^3}{\sum\limits_{k=1}^{25} y_k^3}, \qquad k = 1, 2, ..., 25,\tag{17}$$

$$y_1^4 = o_1^4,\tag{18}$$

where y_1^4 denotes the temperature set-point change u.

3.2. Fuzzy Neural Network Learning Algorithm

Compared with other traditional methods, the BP algorithm has better persistence and predictability. In BP neural network, the error signal is transmitted backward from the output layer to the first layer. Therefore, hierarchical calculation can be reserved and the index function can be defined as:

$$E = \frac{(d - y_1^4)^2}{2},\tag{19}$$

where d is desired signal P_{target}, and y_1^4 denotes P_{total} in this paper.

If the output layer is concerned, we can draw the following Equations (20) from (15) and (16):

$$\begin{aligned}
\delta_1^4 &= -\frac{\partial E}{\partial o_1^4} \\
&= -\frac{\partial E}{\partial P_{total}} * \frac{\partial P_{total}}{\partial u} * \frac{\partial u}{\partial o_1^4} \\
&\approx e * sgn(\frac{\triangle P_{total}}{\triangle u}) * \frac{\partial u}{\partial o_1^4},
\end{aligned}\tag{20}$$

where δ represents the local gradient of the BP neural network. Since $\frac{\partial P_{total}}{\partial u}$ is unknown, we approximately replace it with the symbolic operator $sgn(\frac{\triangle P_{total}}{\triangle u})$, and the learning rate η in the following can compensate the inaccurate calculation form. The fundamentals of the BP neural network is to use the gradient descent method to correct the weights, and Equation (21) is obtained from Equation (20):

$$\triangle w_k = -\frac{\partial E}{\partial o_1^4} * \frac{\partial o_1^4}{\partial w_k} = \delta_1^4 * y_k^3, \tag{21}$$

where $k = 1, 2, ..., 25$.

For the fuzzy rule layer and the membership function layer, the local gradients are denoted as

$$\delta_k^3 = -\frac{\partial E}{\partial o_k^3} = \delta_1^4 * w_k, \tag{22}$$

$$\delta_j^2 = -\frac{\partial E}{\partial o_j^2} = (\sum_k \delta_k^3 * y_i^2) * y_j^2, \quad j = 1, 2, ..., 10, \tag{23}$$

where $j \neq i, k$ ($k = 1, 2, ..., 25$) stands for the node in the third layer, which are connected with the jth node in the second layer. i denotes another node in the second layer that is connected to the kth node in the third layer.

Therefore, the parameter correction value of the input membership function is as follows:

$$\triangle m_{ij} = -\frac{\partial E}{\partial m_{ij}} = \delta_j^2 \frac{2(y_i^1 - m_{ij})}{\sigma_{ij}^2}, \tag{24}$$

$$\triangle \sigma_{ij} = -\frac{\partial E}{\partial \sigma_{ij}} = \delta_j^2 \frac{2(y_i^1 - m_{ij})^2}{\sigma_{ij}^3}. \tag{25}$$

Finally, the correction algorithm of adjustable parameters in fuzzy neural networks is expressed by

$$w_k(n+1) = w_k(n) + \eta_1 \triangle w_k(n) + \\ \alpha_1(w_k(n) - w_k(n-1)), \tag{26}$$

$$m_{ij}(n+1) = m_{ij}(n) + \eta_2 \triangle m_{ij}(n) + \\ \alpha_2(m_{ij}(n) - m_{ij}(n-1)), \tag{27}$$

$$\sigma_{ij}(n+1) = \sigma_{ij}(n) + \eta_3 \triangle \sigma_{ij}(n) + \\ \alpha_3(\sigma_{ij}(n) - \sigma_{ij}(n-1)), \tag{28}$$

where η_1, η_2, η_3 are learning rates for each adjustable parameter, and $\alpha_1, \alpha_2, \alpha_3$ are momentum factors for each adjustable parameter. The well-chosen learning factors and momentum factors can accelerate algorithm convergence, reduce shock, and effectively suppress the local minimum—and their values are limited to $(0, 1)$ interval.

3.3. Optimization of Initial Value of Adjustable Parameters

The learning effect of the fuzzy neural network has a strong dependence on the initial values of the connection weight and membership function. The BP algorithm is suitable for solving the complicated nonlinear problem, but the algorithm usually falls into local minimum, resulting in training failure. In [26], the authors proposed a method for estimating the parameters of dynamic models for induction motor dominating loads. Based on PSO, the method finds the adequate set of parameters that best fit the sampling data from the measurement for a period of time, minimizing the error of the outputs and active and reactive power demands. Hence, a hybrid algorithm uniting the PSO and BP algorithms was proposed to find the optimal initial parameter. The PSO algorithm includes a group of particles, these particles are stochastically distributed in the high-dimensional search space. These particles are group members that are used to find the optimal solution in a high-dimensional search space, and the optimal solution is equal to the possible position for the object. Each particle is a candidate optimal solution in a higher dimensional search space, the optimal direction and speed of individual particles are dependent on the optimal location and optimal speed of the whole particle and its own, and the optimization results are measured by the fitness function. The fitness function is determined by the specific problem. The mathematical description is as follows:

$$V_i = V_i + c_1 * r_1 * (lbest_i - L_i) + c_2 * r_2 * (gbest - L_i), \tag{29}$$

$$L_i = L_i + V_i, \tag{30}$$

where V_i, L_i and $lbest_i$ denote the velocity, location, and historical optimal location of the ith particle, respectively. $gbest$ is the best position of all particles at present. c_1 and c_2 are learning rates, and r_1 and r_2 are two random numbers between 0 and 1.

Next, the root-mean-square error (RMSE) is defined as an indicator used to assess tracking performance, which can be defined as follows:

$$RMSE = \sqrt{\frac{\sum\limits_{k=1}^{N_s} e_k^2}{N_s (P_{target}^{max} - P_{target}^{min})^2}}, \tag{31}$$

where N_s denotes quantity of control cycles, P_{target}^{max} and P_{target}^{min} denote the upper and lower limits of the target signal range, respectively. The flow chart of the optimization algorithm is shown in Figure 5.

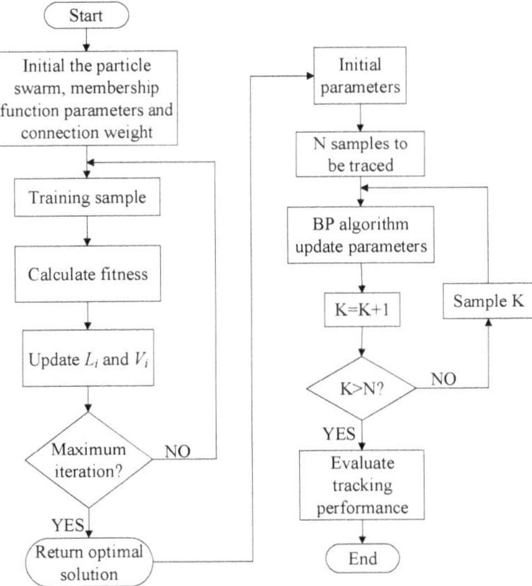

Figure 5. Flow chart of the optimization algorithm.

4. Simulation Results

In the simulations, the HVAC units track the daily automatic generation control (AGC) signal from the PJM (Pennsylvania–New Jersey Maryland Interconnection) electricity markets. Following the parameters of TCLs in the simulation process [7], it is assumed that the average values of R, C, and P in Equation (1) follow a Gaussian distribution with standard deviation 0.1, and the values of R, C, and P are 2 °C/kW, 2 kWh/°C, and 14 kWh, respectively. Assume the loads' initial temperatures are distributed uniformly in the deadband of temperature, T^{set} is 20 °C and T_a is 32 °C. The deadband of temperature $\triangle$ is 0.5 °C, η in Equation (4) is 2.5, and the sampling time interval is 4 s. The tracking errors are used to characterize the performance of the frequency regulation based on the fuzzy neural network control.

Through the iterative solution of the particle swarm optimization algorithm and BP algorithm, the appropriate connection weight coefficient is obtained after the training of 2500 sample dates. The membership functions of error e and error variation de are shown in Figures 6 and 7, respectively. At the beginning of the simulation, we assume that the membership functions of the fuzzy sets of error e and error variation de are Gaussian functions, as shown in Figures 6a and 7a—after training, the membership functions of each fuzzy set of the fuzzy neural network are shown in Figures 6b and 7b.

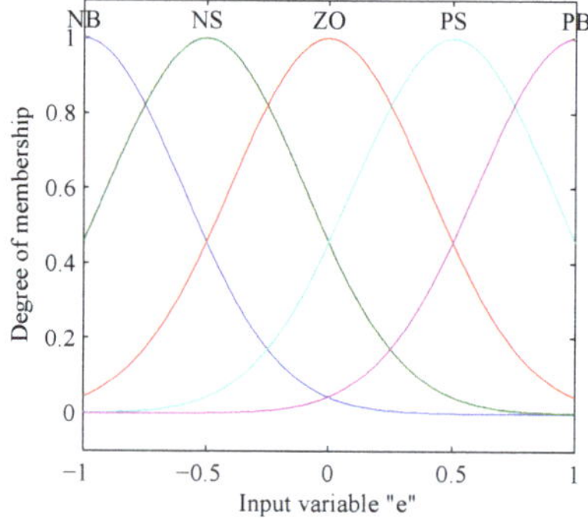

(**a**) The initialized membership function of error e.

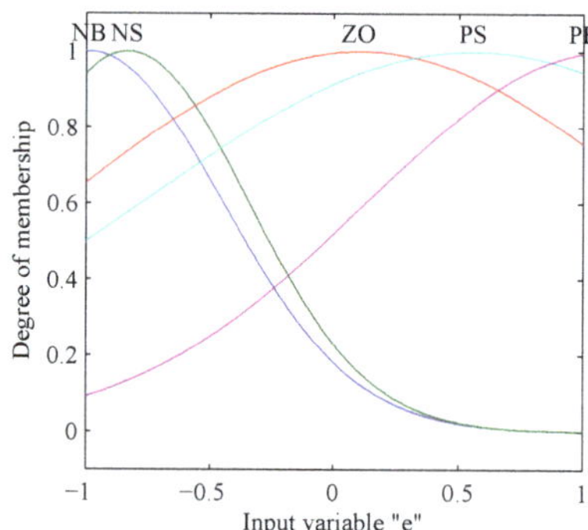

(**b**) The optimized membership function of error e.

Figure 6. The membership function of error e.

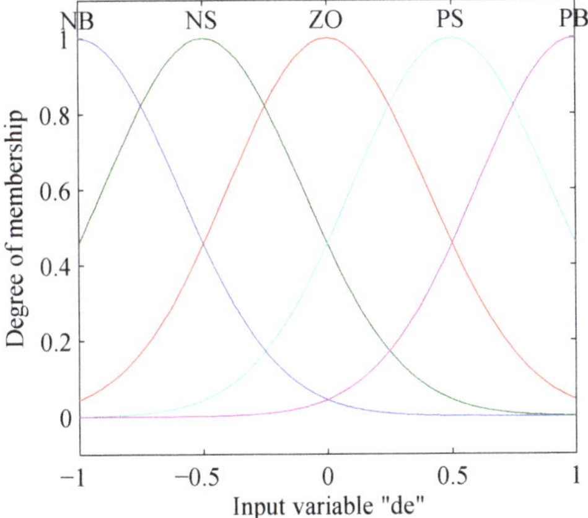

(**a**) The initialized membership function of error variation *de*.

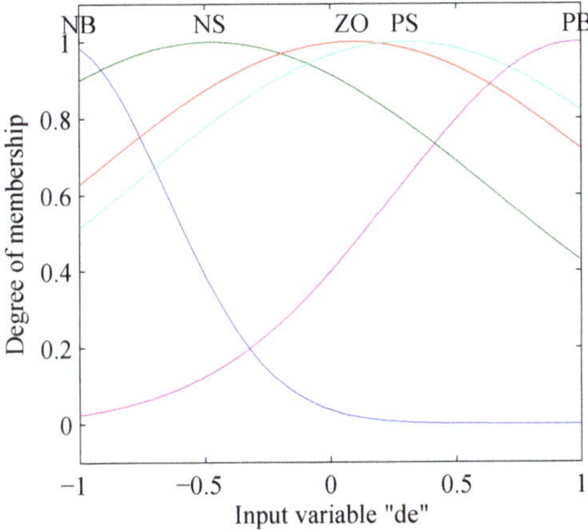

(**b**) The optimized membership function of error variation *de*.

Figure 7. The membership function of error variation *de*.

The influence of the PSO algorithm on the performance of the neural fuzzy network is shown in Figure 8. From the results shown in Figure 8, it can be observed that the RMSE with optimized parameter under PSO algorithm is 2.7003, which is obviously smaller than the RMSE with random parameters. Hence, it is necessary to optimize these parameters by the PSO algorithm.

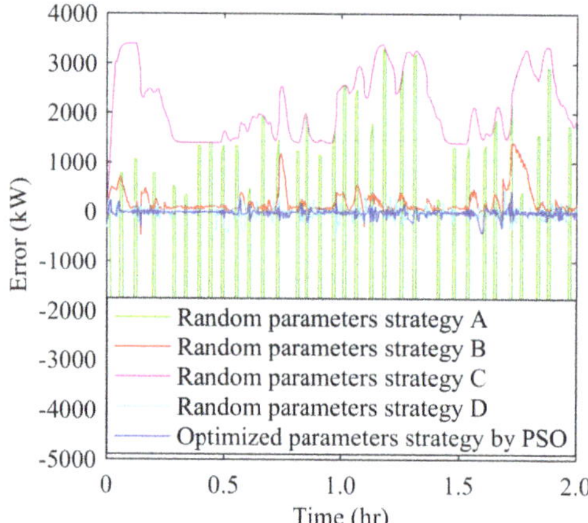

Figure 8. Error comparisons with non-optimized and optimized parameters.

The control performance of the fuzzy neural network is shown in Figure 9, and the changes in temperature set-point are shown in Figure 10. We observe that the aggregated TCLs can track the AGC signal in a power system to provide ancillary service, and the maximum change in temperature set-point is 1.18 °C, which indicates that the control scheme has a minor impact on the thermal comfort of consumers.

Furthermore, several control strategies are compared for the problems of frequency control. Figure 11 shows the comparison results of tracking errors, it is shown that the fuzzy neural network control strategy can better reduce the tracking errors. Table 1 shows the detailed comparison results, which demonstrate that the fuzzy neural network control strategy can better reduce tracking errors with acceptable temperature set-point change. In addition, it can be observed from Table 1 that there are a small number of switching on/off times using the fuzzy neural network control strategy, which indicates that its regulation has a smaller impact on the life of TCLs compared with other control strategies.

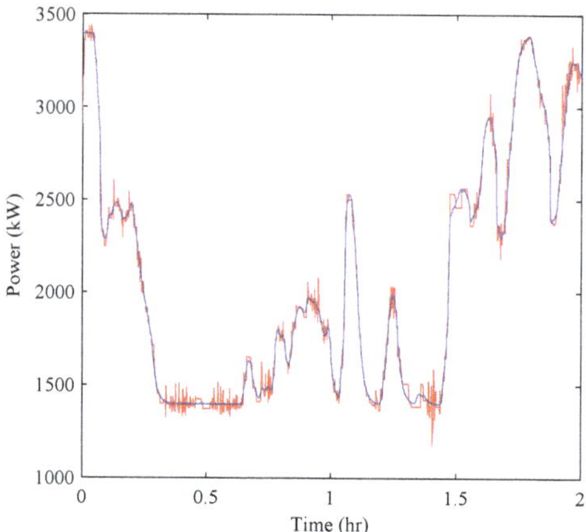

Figure 9. Automatic generation control (AGC) signal tracking.

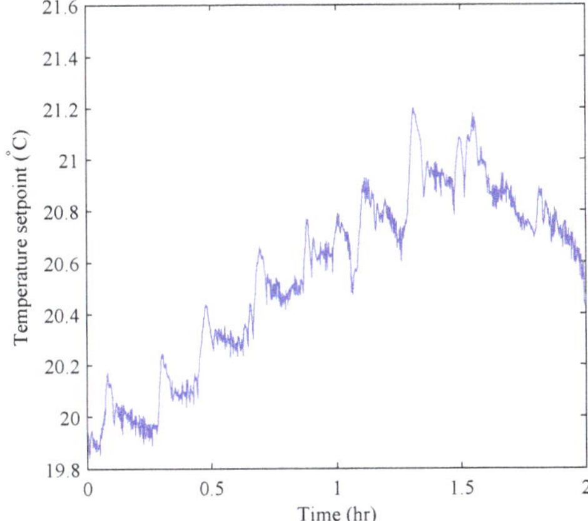

Figure 10. The temperature set-point change.

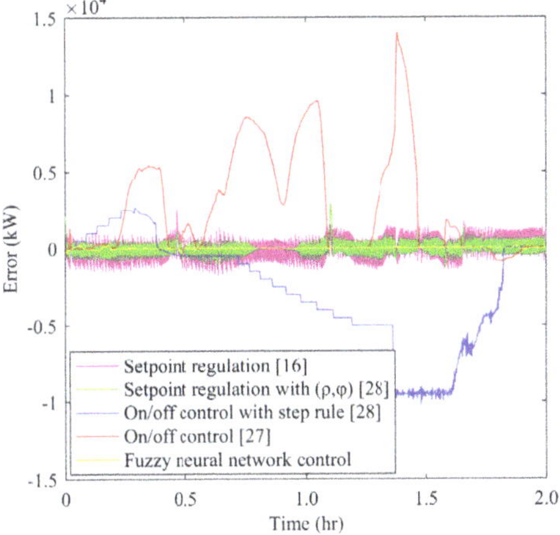

Figure 11. Error comparisons under five control strategies.

Table 1. Comparison results.

Control Strategies	RMSE/%	Set-Point Range/°C	Average On/Off Time
On/off control [27]	19.75	20	62
On/off control with the step rule [28]	14.22	20	45
Set-point regulation [16]	4.01	$19.83 \sim 21.11$	26
Set-point regulation with (ρ,ϕ) [28]	2.80	$19.82 \sim 21.02$	15
Fuzzy neural network control	2.33	$19.83 \sim 21.18$	8

5. Conclusions

In this paper, the frequency regulation method—based on fuzzy neural network control—is proposed to regulate the temperature set-point of TCLs in the ancillary service market. Due to the strong dependence of tracking accuracy on membership functions and connection weight coefficients, the combination of offline hybrid algorithms and online BP algorithms can better optimize the control parameters. Finally, the conclusion obtained from the simulation results is that the fuzzy neural network control has prominent advantages of tracking accuracy without modeling the aggregated TCLs.

Author Contributions: K.M. contributed the idea; Z.Q. wrote the paper; C.X. conceived and designed the experiments; Z.J. provided the software.

Funding: This work was supported by the National Key Research and Development Program of China under Grant 2016YFF0200105 and National Natural Science Foundation of China under Grant 61573303.

Conflicts of Interest: The authors declare no conflict of interest. The founding sponsors had no role in the design of the study; in the collection, analyses, or interpretation of data; in the writing of the manuscript, and in the decision to publish the results.

References

1. Ma, K.; Liu, X.; Liu, Z.; Chen, C.; Liang, H.; Guan, X. Cooperative relaying strategies for smart grid communications: bargaining models and solutions. *IEEE Internet Thing. J.* **2017**, *4* 2315–2325. [CrossRef]
2. Hashemi-Dezaki, H.; Askarian-Abyaneh, H.; Haeri-Khiavi, H. Impacts of direct cyber-power interdependencies on smart grid reliability under various penetration levels of microturbine/wind/solar distributed generations. *IET Gener. Trans. Distrib.* **2016**, *10*, 928–937. [CrossRef]
3. Wen, G.; Hu, G.; Hu, J.; Shi, X.; Chen, G. Frequency regulation of source-grid-load systems: A compound control strategy. *IEEE Trans. Ind. Inf.* **2016**, *12*, 69–78.
4. Wang, D.; Jia, H.; Wang, C.; Lu, N. Performance evaluation of controlling thermostatically controlled appliances as virtual generators using comfort-constrained state-queueing models. *IET Gener. Trans. Distrib.* **2009**, *8*, 591–599. [CrossRef]
5. Lu, N.; Vanouni, M. Passive energy storage using distributed electric loads with thermal storage. *J. Mod. Power Syst. Clean Energy* **2013**, *1*, 264–274. [CrossRef]
6. Ma, K.; Hu, G.Q.; Spanos, C.J. Energy management considering load operations and forecast errors with application to HVAC systems. *IEEE Trans. Smart Grid* **2018**, *9*, 605–614. [CrossRef]
7. Bashash, S.; Fathy, H.K. Modeling and control insights into demand-side energy management through setpoint control of thermostatic loads. In Proceedings of the American Control Conference (ACC), San Francisco, CA, USA, 29 June–1 July 2011; pp. 4546–4553.
8. Ledva, G.S.; Vrettos, E.; Mastellone, S.; Andersson, G.; Mathieu, J.L. Managing communication delays and model error in demand response for frequency regulation. *IEEE Trans. Power Syst.* **2018**, *33*, 1299–1308. [CrossRef]
9. Zhou, K.; Cai, L. A dynamic water-filling method for real-time HVAC load control based on model predictive control. *IEEE Trans. Power Syst.* **2013**, *30*, 1405–1414. [CrossRef]
10. Mahdavi, N.; Braslavsky, J.H.; Seron, M.M.; West, S.R. Model predictive control of distributed air-conditioning loads to compensate fluctuations in solar power. *IEEE Trans. Smart Grid* **2017**, *8*, 3055–3065. [CrossRef]
11. Liu, M.; Shi, Y.; Liu, X. Distributed MPC of aggregated heterogeneous thermostatically controlled loads in smart grid. *IEEE Trans. Ind. Electron.* **2016**, *63*, 1120–1129. [CrossRef]
12. Callaway, D.S. Tapping the energy storage potential in electric loads to deliver load following and regulation, with application to wind energy. *Energy Convers. Manag.* **2009**, *50*, 1389–1400. [CrossRef]
13. Kundu, S.; Sinitsyn, N.; Backhaus, S.; Hiskens, I. Modeling and control of thermostatically controlled loads. In Proceedings of the Power Systems Computation Conference, Stokholm, Sweden, 22–26 August 2011. Available online: https://arxiv.org/abs/1101.2157 (accessed on 25 June 2019).
14. Perfumo, C.; Braslavsky, J.H.; Ward, J.K. Model-Based Estimation of Energy Savings in Load Control Events for Thermostatically Controlled Loads. *IEEE Trans. Smart Grid* **2014**, *5*, 1410–1420. [CrossRef]

15. Trovato, V.; Tindemans, S.H.; Strbac, G. Leaky storage model for optimal multi-service allocation of thermostatic loads. *IET Gener. Trans. Distrib.* **2015**, *10*, 585–593. [CrossRef]

16. Ma, K.; Yuan, C.; Yang, J.; Liu, Z.; Guan, X. Switched Control Strategies of Aggregated Commercial HVAC Systems for Demand Response in Smart Grids. *Energies* **2017**, *10*, 953. [CrossRef]

17. Ni, Z.; Wang, M. Research on the Fuzzy Neural Network PID Control of Load Simulator Based on Friction Torque Compensation. In Proceedings of the Sixth International Conference on Intelligent Human-Machine Systems and Cybernetics, Hangzhou, China, 26–27 August 2014.

18. Garcia-Valle, R.; Silva, L.C.P.D.; Xu, Z.; Ostergaard, J.J. Smart demand for improving short-term voltage control on distribution networks. *IET Gener. Trans. Distrib.* **2009**, *3*, 724–732. [CrossRef]

19. Wai, C.H.; Beaudin, M.; Zareipour, H.; Schellenberg, A.; Lu, N. Cooling devices in demand response: A comparison of control methods. *IEEE Trans. Smart Grid* **2014**, *6*, 249–260. [CrossRef]

20. Ma, K.; Hu, G.Q.; Spanos, C.J. Distributed Energy Consumption Control via Real-Time Pricing Feedback in Smart Grid. *IEEE Trans. Control Syst. Technol.* **2014**, *22*, 1907–1914. [CrossRef]

21. Ma, K.; Wang, C.; Yang, J.; Hua, C.; Guan, X. Pricing mechanism with noncooperative game and revenue sharing contract in electricity market. *IEEE Trans. Cybern.* **2017**, *49*, 97–106. [CrossRef]

22. Lin, F.J.; Shieh, P.H.; Hung, Y.C. An intelligent control for linear ultrasonic motor using interval type-2 fuzzy neural network. *IET Electr. Power Appl.* **2018**, *2*, 32–41. [CrossRef]

23. Chen, Y.C.; Teng, C.C. A model reference control structure using a fuzzy neural network. *Fuzzy Sets Syst.* **1995**, *73*, 291–312. [CrossRef]

24. Kong, S.G.; Kosko, B. Adaptive fuzzy systems for backing up a truck-and-trailer. *IEEE Trans. Neural Netw.* **1992**, *3*, 211–223. [CrossRef]

25. Berenji, H.R.; Khedkar, P. Learning and tuning fuzzy logic controllers through reinforcements. *IEEE Trans. Neural Netw.* **1992**, *3*, 724–740. [CrossRef] [PubMed]

26. Kim, Y.; Song, H.; Lee, B. Identification of Dynamic Load Model Parameters Using Particle Swarm Optimization. *Int. J. Fuzzy Log. Intell. Syst.* **2010**, *10*, 128–133. [CrossRef]

27. Hao, H.; Sanandaji, B.M.; Poolla, K.; Vincent, T.L. Aggregate flexibility of thermostatically controlled loads. *IEEE Trans. Power Syst.* **2015**, *30*, 189–198. [CrossRef]

28. Ma, K.; Yuan, C.; Liu, Z.; Yang, J.; Guan, X. Hybrid control of aggregated thermostatically controlled loads: Step rule, parameter optimisation, parallel and cascade structures. *IET Gener. Trans. Distrib.* **2016**, *10*, 4149–4157. [CrossRef]

Article

Load Disaggregation Using Microscopic Power Features and Pattern Recognition

Wesley Angelino de Souza [1], Fernando Deluno Garcia [2], Fernando Pinhabel Marafão [2,*],
Luiz Carlos Pereira da Silva [3] and Marcelo Godoy Simões [4,*]

[1] Department of Computer Science, Federal University of São Carlos (UFSCar), Sorocaba SP 18052-780, Brazil
[2] Institute of Science and Technology of Sorocaba, São Paulo State University (UNESP), Sorocaba SP 18087-180, Brazil
[3] School of Electrical and Computer Engineering (FEEC), University of Campinas (UNICAMP), Campinas SP 13083-852, Brazil
[4] Department of Electrical Engineering, Colorado School of Mines, Golden, CO 80401, USA
* Correspondence: fernando.marafao@unesp.br (F.P.M.); msimoes@mines.edu (M.G.S.)

Received: 29 May 2019; Accepted: 5 July 2019; Published: 10 July 2019

Abstract: A new generation of smart meters are called cognitive meters, which are essentially based on Artificial Intelligence (AI) and load disaggregation methods for Non-Intrusive Load Monitoring (NILM). Thus, modern NILM may recognize appliances connected to the grid during certain periods, while providing much more information than the traditional monthly consumption. Therefore, this article presents a new load disaggregation methodology with microscopic characteristics collected from current and voltage waveforms. Initially, the novel NILM algorithm—called the Power Signature Blob (PSB)—makes use of a state machine to detect when the appliance has been turned on or off. Then, machine learning is used to identify the appliance, for which attributes are extracted from the Conservative Power Theory (CPT), a contemporary power theory that enables comprehensive load modeling. Finally, considering simulation and experimental results, this paper shows that the new method is able to achieve 95% accuracy considering the applied data set.

Keywords: load disaggregation; artificial intelligence; cognitive meters; machine learning; state machine; NILM

---

1. Introduction

Eectricity metering has undergone significant technological progress over the last 30 years, from electromechanical to electronic metering. An essential stage of this evolution arose with Automatic Meter Reading (AMR) [1], which includes the following main features [2,3]:

- Improved accuracy in terms of energy readings when compared to the preceding electromechanical meters, in which the measuring errors were quite susceptive and dependent on human operators;
- Automatic and frequent meter readings;
- Customer support improvement;
- More consumption information;
- Support to hourly price charges.

Moreover, the AMR was the basis for the succeeding evolution called Advanced Metering Infrastructure (AMI), which includes the following characteristics [4–6]:

- Bidirectional communication;
- A power consumption monitoring system;

- Advanced and accurate sensors;
- The embedded system is responsible for data collection and manages the required information between meter and utility.

AMI is related to the entire metering infrastructure and 'smart meter' is the popular name for the power metering device in this infrastructure. The term "smart" makes sense in the data processing approach, that is, the meter might process input data (voltage and current), transforming it into useful output information (e.g., energy consumption, power quality indicators, efficiency, and others). However, the concept of "smart" is not usually well defined, since most of the smart meters on the market do not have any smart functionality. Thus, there is a current demand for innovative and intelligent techniques to provide different sorts of information to utilities and consumers, improving their knowledge about energy use, efficiency, costs, consequently improving energy management. Therefore, researchers all over the world are proposing new tools and methods to provide further information about energy consumption [7,8], as well as proposing innovative ways to save energy [9–12].

In this context, a new generation of smart meters are called cognitive meters, which propose to use artificial intelligence and load disaggregation methods, also known as Non-Intrusive Load Monitoring (NILM). They recognize appliances connected to the grid during certain periods, while providing much more information to consumers than the traditional monthly consumption. Consequently, consumers' operations can be inspected to provide them with detailed information about their electrical consumption [13] so they can make better decisions concerning saving electricity, as well as implementing energy management systems for automatic generation/consumption regulation. This is certainly a meaningful advance concerning the relationship between utilities and consumers [11–13]. Moreover, "c-meters" can provide advanced functionalities, such as detailed recommendation of when to use some particular appliances (according to statistical behavior, real-time consumption or hourly-energy price) and they can also suggest some tip to save electricity over weeks, months and years.

Hart [14] initially introduced the NILM method, considering active power levels and distributing them into individual appliance data. With such a type of cognition, the consumer profile can be mapped and by using artificial intelligence techniques, new methodologies can be proposed for modern smart meters [7,8,10,14–22]. However, although NILM is quite a good approach to detect home appliances, it does not always present a reasonable accuracy, demanding other signal analysis or AI to improve detection accuracy.

Therefore, this paper introduces the Power Signature Blob (PSB)—a novel methodology that correlates a hybrid load disaggregation technique, which is based on feature extraction from current and voltage waveforms with power signatures. In this context, a predefined threshold level is compared to the active power variation, in addition to the power signatures. The procedure uses the difference between the actual active power and the last active power value used to define the step level direction—when the appliance is turned on or turned off. Considering that every step level detection is a new event, the novel NILM method calculates the proper features to classify the load and then, applies machine learning for appliance recognition. For classification, the NILM dataset from [15] was used, including instances of 35 appliances (In this paper, the term sample is defined as a signal acquisition from the voltage and current waveforms, and the term instance represents a dataset sample with the respective features.). Each appliance instance comprises active power and other power components (power factor, reactivity factor, and distortion factor). The features to classify and from the dataset are calculated using a contemporary power theory called Conservative Power Theory (CPT) [23,24]. Hence, the novelty of this paper is a new state-machine NILM that uses and acknowledges a dataset of 35 appliances with features based on power components by the CPT.

The next section discusses the main concept of load disaggregation and different techniques from the literature. Afterwards, the PSB method, using the power decompositions from CPT [23–25], is presented. Finally, simulations and experimental results depict the performance of the proposed approach.

2. NILM State-of-the-Art Review

Considering current and voltage acquisition and processing, the literature on NILM can be divided into two main categories—"high-frequency" and "low-frequency". The low-frequency is the category in which the features are extracted at 1 kHz or less. The high-frequency is the category in which features are extracted at kHz or MHz [15,19]. Table 1 presents a list of NILM studies and the features extracted from the point of view of the attributes.

Table 1. Summary and categorization of some Non–Intrusive Load Monitoring (NILM) techniques—relevant works and their main features.

Sampling	Signal Sampling Technique	Features (Attributes)	References
Low-frequency (Macroscopic)	Active/reactive power signature and variation	P, Q	[14,26–30]
	Active power signature and variation	P	[31–34]
	Power signature and power indicators	$P, I_{RMS}, U_{RMS}, I_{MAX}, U_{MAX}, PF(\lambda)$	[35,36]
	Power signature and power quality indicators	$P, PF(\lambda)$	[37]
	Load on or off probability	$P_{load}^{ON}, P_{load}^{OFF}$	[38]
	Temporal discrete power pulses	$P_{pulse}(t)$	[39]
High-frequency (Microscopic)	Harmonic decomposition	$P, Q, H_{1,3,5,\cdots,N}$	[40–43]
	Power signature and harmonic decomposition	P, THD	[7,17]
	Power signatures and Power Theory	P, Q, S or $A, H, I_f, CPT_{components}$	[15,44–46]
	Wavelets	W_{coef}, W_e	[19,47–53]
	Voltage and current trajectory	VI_{traj}	[18,54–56]
	Harmonic noise in power transients	$H_f, 36 \le f \le 500Hz$	[57,58]
	Combination of independent features	$VI_{traj}, i(t), p(t)$ and others	[17]

Note: P: active power; Q: reactive power; I_{RMS}: effective current; U_{RMS}: effective voltage; U_{MAX}: maximum voltage; I_{MAX}: maximum current; $PF(\lambda)$: power factor; P_{load}^{ON}: load on probability; P_{load}^{OFF}: load off probability; $P_{pulse}(t)$: time power pulse; $H_{1,3,5,\cdots,N}$: nth harmonic components ($n = 1, 3, ..., N$); THD: voltage or current total harmonic distortion; S or A: apparent power; H: harmonic power; I_f: inactive current; $CPT_{components}$: CPT power components; W_{coef}: wavelet coefficients; W_e: wavelet equivalent coefficients; VI_{traj}: voltage vs current trajectory; H_f: non-fundamental power components; $i(t)$: instantaneous current; $p(t)$: instantaneous power.

In 1992, Hart [14] did some pioneering work on load disaggregation, in which he defined Nonintrusive Appliance Load Monitoring (NALM)—NILM is a derivative term from NALM. Such work showed that it is possible to separate power consumption by appliances observing the collective power consumption. To do this, it is necessary to discover the power behavior of each load or appliance. For a long time, this research did not draw attention due to digital device limitations for embedded algorithms. At the same time, Sultanem [40] proposed a different algorithm to Hart's work and used the PQ trajectory with harmonic decomposition.

Considering the P-Q (active and reactive power) analysis, Drenker and Kader [26] validated their NILM method with six loads that operate in steady mode, obtaining an accuracy of around 95%. Cole and Albicki [27] studied the steady-state loads and some loads with slow power changing (such as heat pump compressors) and they considered the total consumption of each load as geometrical shape forms. Norford and Leeb [28] also detected transient status of some loads and created an approximation for these transient statuses. Biansoongnern and Plungklang [29] created a NILM method with 90% of accuracy when testing air conditioning and refrigerators. Using deep learning to detect operational load changes, Xiao and Cheng proposed a method and validated it using the Reference Energy Disaggregation Dataset (REDD) [59].

Powers et al. [31] applied a rule-based algorithm to detect loads with high consumption, such as air conditioning, water heaters and electric space heaters. Likewise, rule-based algorithms were proposed by Farinaccio and Zmeureanu [32] to detect power load behavior, as well as by Marceau and Zmeureanu [33] with the indication of 90% accuracy. With regards to genetic algorithms, Baranski and Voss [34] proposed their employment to detect patterns based on the use of loads frequency.

Ruzzelli et al. [35] created a dataset with low-frequency features of voltages and currents (I_{RMS}, V_{RMS}, I_{MAX}, V_{MAX}) and the power factor (PF) to describe load behavior, and with the P-Q analysis they carried out load disaggregation. Kelly and Knottenbelt [36] used a deep neural network for load disaggregation, as well as the UK Domestic Appliance-Level Electricity (UK-DALE) dataset [60], achieving an excellent performance for that dataset. Figueiredo et al. [37] used the load step changes in active power and the PF to create a dataset and concluded that there is a need to extract other attributes that can better detail the loads, especially those that have the same power and the same behavior as the equivalent circuit.

Kim et al. [38] combined frequency independent features, such as the distribution probability of ON/OFF duration, frequency of appliance usage and the correlation between the usage of various appliances with the active power feature, achieving between 64% to 99.8% accuracy in terms of load disaggregation. Koutitas and Tassiulas [39] replaced time-series power analysis for a set of discrete pulses. They created features based on pulses, such as variance, spike, slope, periodicity, multi-state, and sequence of operation. They reached an accuracy of 85%.

Sultanem [40] is the pioneer of the high-frequency use on NILM applications. Srinivasan et al. [41] used machine learning to recognize the harmonic signatures of 8 loads, and they obtained an accuracy of 99% or more for the load disaggregation. Laughman et al. [42] used the P-Q analysis for similar loads and increased the 3rd harmonic to distinguish them. Bouhouras et al. [43] used harmonic components to create a dataset for load disaggregation with stand-alone loads or combined loads, and the accuracy was between 85% and 95%.

Dong et al. [7] adopted Total Harmonic Distortion (THD) of current waveforms and P-Q analysis for load discrimination, and they used some pulse-based features. Lin et al. [17] created a NILM using features including the THD, P-Q analysis, voltage-current trajectory, current indicators and quadratic programming. The accuracy of this work is generally more than 90%.

Using power theory concepts, Teshome et al. [44] proposed a NILM with components of active, reactive, apparent power, and nonactive currents. Nguyen et al. [45] created a NILM method based on active, reactive and apparent power and used a decision tree (DT) to disaggregate five loads, achieving more than 98.8% accuracy. Huang et al. [46] pointed out that the application of power theories can be a useful tool for load disaggregation, especially for loads that have the same value of active power. In 2003, Tenti and Mattavelli [23,24] proposed the Conservative Power Theory, which allows load modelling in terms of power components. This work was the basis for the load characterization in the NILM dataset proposed by Souza et al. [15,25].

Considering the Continuous Wavelet Transform (CWT), Chan et al. [47] applied it up to the 4th level for the loads, and used *Daubechies* as mother wavelets, achieving an accuracy of 70%. Su et al. [48], and Duarte et al. [49] compared the Fast Fourier Transform (FFT) with CWT, pointing out the advantages of the CWT during load transients and recommended using CWT for feature extraction for load disaggregation. Chang et al. [52,53] implemented a NILM using active and reactive power with CWT. The authors extracted load features from five filters and applied a genetic algorithm for load identification. They reached almost 100% accuracy but the studies were carried out considering situations of significant discrepancy in power levels.

Gray and Morsi [50] used time-consuming energy to obtain CWT decomposition with Daubechies mother wavelets. The authors presented results comparing the accuracy of applying each order of the Daubechies order and concluded that the higher the Daubechie, the greater the accuracy. Tabatabaei [51] did a similar study, but used power characteristics to create classificatory features and obtained accuracy at around 85%. Gillis et al. [19] proposed a new CWT for NILM applications, obtaining around 94% accuracy for four connected loads at the same time.

Hassan et al. [54] created a NILM method based on V-I trajectory, which used wave-shape features along with the REDD dataset. Similarly, V-I trajectories were mapped to a grid of cells (as a matrix) by Du et al. [18], having a binary value assigned to each of them. On the other hand, 83% of accuracy was achieved by Gao et al. [55] by converting V-I trajectory into a binary image, while using combined features, and considering 11 appliances. Finally, convolutional neural networks were proposed to be used with V-I trajectory by Baets et al. [56], reaching 77.6% of accuracy for the PLAID dataset [61], and 75.46% for the WHITED dataset [62]. Voltage harmonic (FFT) noise has also been used by Patel et al. [57,58], taking into account noise and electromagnetic interference in the range of 36 to 500 kHz. Nonetheless, such a study only highlighted the types of equipment that present multiple operational stages.

To summarize, there are many other studies regarding NILM with different methodologies, feature extraction, different appliances in the validation and different load disaggregation algorithms. Nevertheless, in Teshome et al. [44], the authors indicate the importance of modern power theories and the lack of these elegant circuit analyses to improve the NILM systems. One of these elegant power theories pointed out in Teshome et al. [44] was the CPT. Thus, such a modern power theory is applied in this work to improve the load disaggregation and present a novel NILM technique.

Accordingly, the next section presents the PSB, a new NILM methodology based on a state machine, which analyzes the active power signature (a low-frequency feature), and on the event detection, which finds features from the CPT and triggers the machine learning algorithm that uses the high-frequency attribute dataset proposed by Souza et al. [15].

3. The Power Signature Blob Method

3.1. Dataset with the Microscopic Features Extraction

In Souza et al. [15] some techniques for appliance disaggregation were evaluated, and the feasibility of identifying home appliances using pattern recognition algorithms was shown. Two pattern recognition algorithms achieved significant results: Optimum-Path Forest (OPF) [63] and K-Nearest Neighbor (KNN) [64] and the KNN (with $K = 1$) was chosen because of its lower computational time. The voltage and current waveforms from several appliances were measured and decomposed in power components using CPT [23,24]. The CPT allows splitting the power into active, reactive, unbalance and residual parts. These power components help to interpret an appliance as an equivalent circuit, as shown in Figure 1, where v_m is the phase voltage, the current i_{Gm} coincides with the active current, i_{Lm} coincides with the reactive current, and the current source j_m coincides with the void current. G_m is the equivalent phase conductance and could be represented as a resistance, L_m is the equivalent phase inductance and could be represented as an inductor. All the mathematical background of the equivalent circuit can be found in Reference [65].

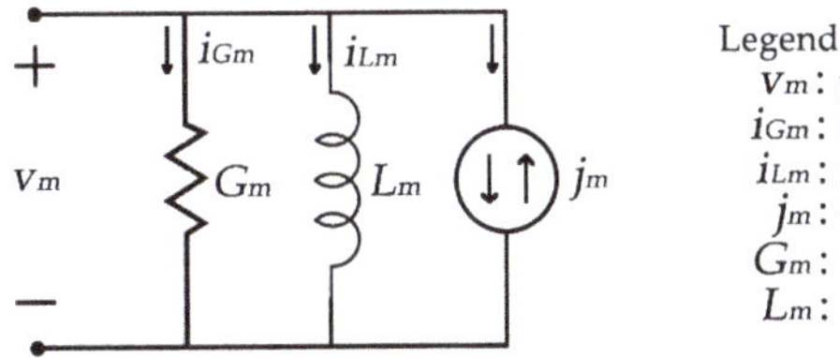

Figure 1. Load equivalent circuit by Conservative Power Theory (CPT).

Using the CPT power decomposition, Souza et al. [15] created a dataset of 35 home appliances such as irons, microwaves, refrigerators, washing machines, lamps and others. Each appliance features (CPT active power, power factor, reactivity factor, and nonlinearity factor [25]) refer to a set dimension and each collected instance as a point in the multidimensional space. The pattern recognition algorithm uses this dataset for the classification purpose of the appliance.

Figure 2 shows the Voronoi diagram concerning the 1NN results for the appliance dataset of [15], where each class number represents an appliance. It shows three Voronoi diagrams for the four attributes (P—Active power, PF—power factor, QF—reactivity factor, and VF—nonlinearity factor) from the dataset, presented in two dimensions to help visualize the decision boundaries.

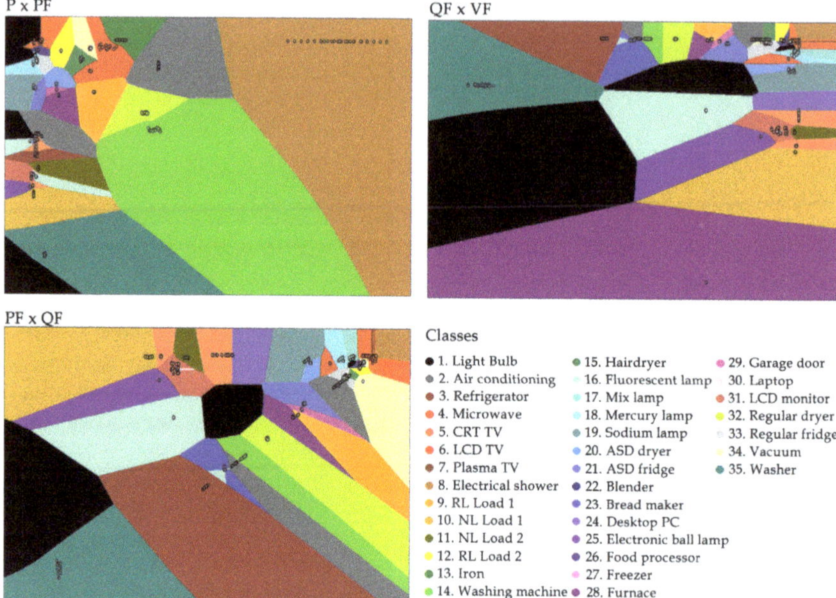

Figure 2. K-Nearest Neighbor (KNN) decision boundaries of the CPT appliance dataset.

The supervised classification algorithm identifies which appliances consume electricity, according to CPT power terms calculation. However, it is likely that some false positives could be detected (when the appliance is identified mistakenly due to the similarity with others, for example, a 100 W bulb lamp and a 100 W LCD TV with high power factor). Another potential issue concerns some appliances with multiple power stages during the power operation, such as washing machines (with washing, spinning and rinsing). This occurs because the classifier needs to observe various levels of "ON" and "OFF" and could not perform the correct classification. Some methodologies [7,8,10,16,22,37,42,66] were created to solve the multiple power steps problem, and both observe the appliance behavior during time operation. The load power signature is relevant because some appliances do not operate in steady-state, and the method needs to disaggregate with high accuracy. Thus, in Reference [15] there was a microscopic feature dataset for the load classification, but it is required to create a method to observe the appliance power behavior before using the appliance classification.

3.2. Load Power Signatures

As initially pointed out in Reference [14], each appliance has a power signature that is not necessarily a steady-state power, thus Figure 3 presents different appliances' power behaviors, and the appliances can be classified using power signatures [7,10,14,16,22], as shown in Figure 4.

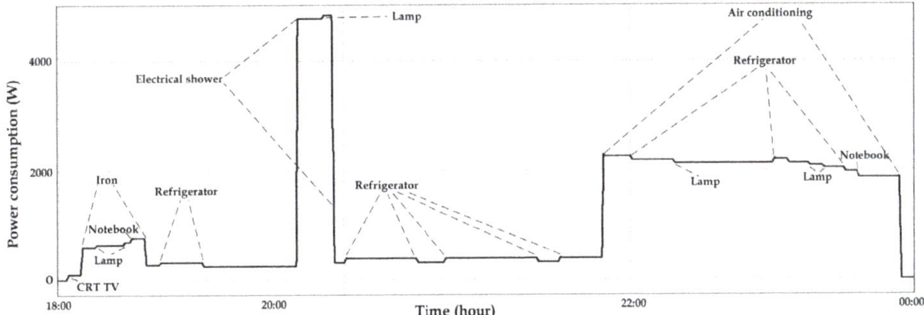

Figure 3. Individual appliance events.

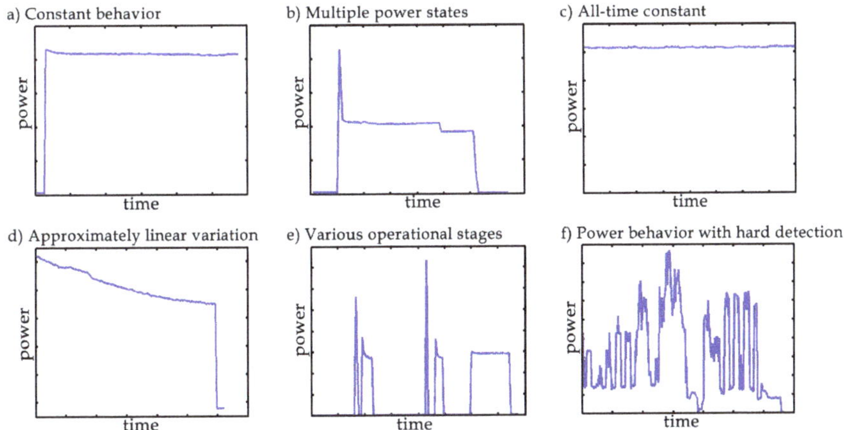

Figure 4. Different appliance power signatures. (**a**) Constant behavior; (**b**) Multiple power states; (**c**) All-time constant; (**d**) Approximately linear variation; (**e**) Various operational stages; and (**f**) Power behavior with hard detection.

From Figure 4, an appliance can have a power signature with:

(a) Constant behavior: In this case, the appliance has steady power behavior over time. It has a high probability of being an appliance with the resistive equivalent circuit, or the appliance is in steady state;

(b) Multiple power states: The appliance has a high-power peak when starting the operation. This characteristic could be linked to a starting engine, and after such a power step, it presents some power variations without turning off. This case can be related to a motorized appliance such as a washing machine;

(c) All-time constant: in this case, the appliance is always connected as a standby mode device;

(d) Approximately linear variation: in this case, there is a linear power variation over time. This variation corresponds to a transitional period until stabilization, such as by temperature (such as an iron) or gas (such as mercury lamp);

(e) Various operational stages: Some appliances have several power characteristics over time, which can be inductive, resistive or non-linear, etc. Besides, the power could also be switched on and off during the operation. For example, a clothes dryer has a motor that can rotate at different speeds and can also have a heating system to facilitate the drying process;

(f) Power behaviors with hard sequence detection: This type of appliance is usually electronic, and there are several fast operation stages, making it difficult to recognize the power signature over time. The noises from current and voltage sensors are also aggregated into this power signature category. The printer is an example of this type of load, which has some power steps that vary and switch very quickly.

Hence, considering the possibility of such different appliances' signatures, it would be important to have a preliminary filter before using any appliance recognition technique, so as to increase the disaggregation accuracy. Thus, the next section presents the proposed approach, which uses CPT power terms and NILM techniques to detect the power signatures, before using the appliance classification method by means of the KNN algorithm.

The power signature behavior was aggregated into the 35 appliances dataset, as can be seen in Table 2. This characteristic is not used as features into the pattern recognition algorithm [15], but it is used to filter and increase accuracy in load detection.

Table 2. Household appliance dataset.

Order (Class)	Load	Event Type (Based on Figure 4)
1	Light bulb	a
2	Air conditioning	b
3	Refrigerator	e
4	Microwave	b
5	CRT TV	a
6	LCD TV	a
7	Plasma TV	a
8	Electrical shower	a
9	RL Load 1	a
10	NL Load 1	a
11	NL Load 2	a
12	RL Load 2	a
13	Iron	d
14	Washing machine	b
15	Hairdryer	b
16	Fluorescent lamp	a
17	Mix lamp	d
18	Mercury lamp	d
19	Sodium lamp	d
20	ASD Dryer	e
21	ASD Fridge	e
22	Blender	b
23	Bread maker	a
24	Desktop PC	e
25	Electronic ball lamp	d
26	Food processor	b
27	Freezer	e
28	Furnace	a
29	Garage door	a or e
30	Laptop	e
31	LCD monitor	a
32	Regular dryer	b
33	Regular fridge	e
34	Vacuum	e
35	Washer	b

3.3. The PSB Technique

The appliance dataset helps to detect one appliance by execution but might result in a problem, since more than one appliance may be ON during a certain period of time. In such a case, the active power would contain the aggregated value. Therefore, in order to decompose the consumption of each individual load, it is necessary to know how many loads are operating at that time.

Thus, considering the existence of an algorithm to classify the appliances that generate a power event, the set of instances and the appliance class create a new key-pair value. Therefore, a set of instances may contain several classes associated, and the aggregated set of instances can be decomposed considering the type of appliances.

To obtain a more accurate value of load energy consumption, the PSB takes the mean value of each block associated with each class to evaluate the mean active power during that period. Blocks that contain more than one active appliance should use the historical average value from each of them. This procedure is called disaggregation by blocks.

The load identification algorithm uses classification attributes extracted from the difference between two scenarios: before and after an appliance is turned ON. Hence, the step is just a divider between stable states. This guarantees that classifiers represent the appliances altogether, mitigating the noise generated in the transition state.

Subtracting the waveforms of these steps creates an approximation of the load waveform. Then, the CPT algorithm generates the attributes by processing the waveform. Adopting this approach, four features represent the appliances: active power, power factor, reactive factor, and non-linearity factor. Using these elements as attributes in a four-dimensional space, each load will result in a cluster. Therefore, the classification algorithms, such as KNN, can be used to set the frontiers of each load in the space.

The diagram from Figure 5 shows the state-machine algorithm with the appliance disaggregation dataset from [15]. The existence of appliance power signatures can be observed, which filter and can make the appliance disaggregation more accurate. Moreover, the methodology has algorithms for handling the ON and OFF events, presented in the diagrams from Figures 6 and 7. The methodology stores the fifteen previous cycles (0.25 s of total samples) of voltage and current waveforms to detect the appliance events if there are two or more appliances turned ON.

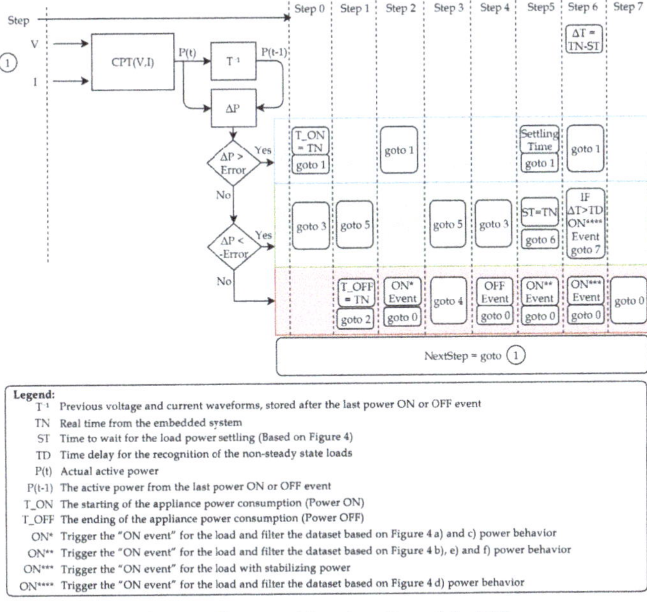

Figure 5. State-machine algorithm of the PSB.

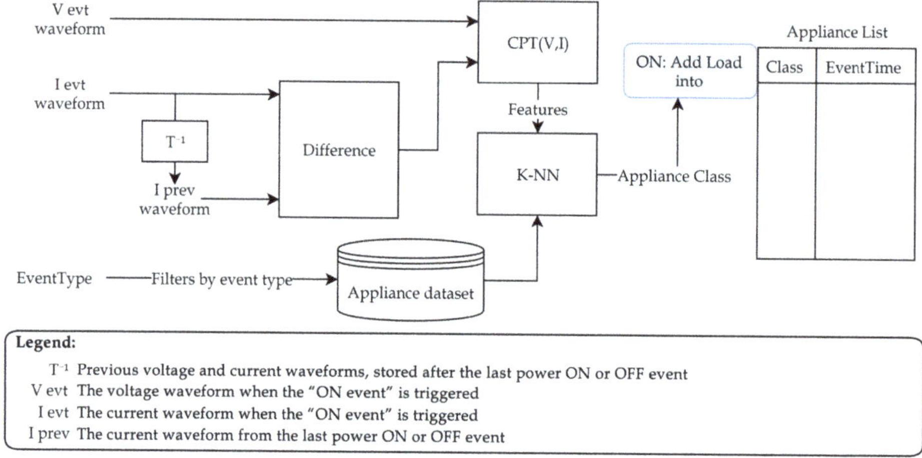

Figure 6. Event ON Trigger.

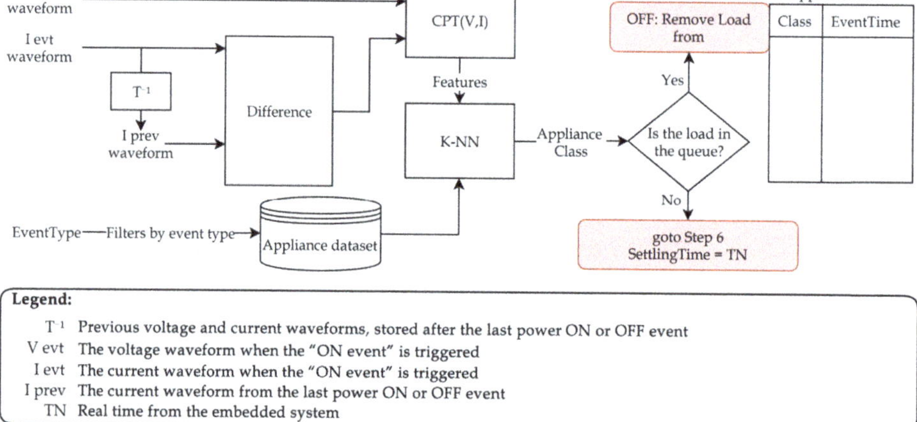

Figure 7. Event OFF Trigger.

4. Validation Results

In this paper, the PSB evaluation was based on simulations and experimental results, as depicted in the next subsections.

4.1. Simulation Results

Table 3 presents the daily appliance schedule for the simulation according to user behavior in the residence.

Table 3. Simulation: Appliance Schedule Runtime.

Appliance	"Turn on" Time	Total Time
Electrical shower	07:15 and 19:40	00:20 and 00:15
Air conditioner	22:00	08:00
CRT TV	12:15 and 18:15	01:00 and 05:00
Refrigerator	All the day	00:30 each cycle
Iron	18:30	00:15
Lamp1 (bulb 100 W)	18:30	05:30
Lamp2 (bulb 60 W)	19:00	04:30
Notebook	19:00	04:00
Microwave	07:45	00:05
Washing Machine	13:00 (Saturday)	02:00

Therefore, an electrical circuit model was created in PSIM software in order to simulate the power behavior of a residence during an entire day, turning on each appliance according to the scheduled runtime of Table 3.

The simulation collects 256 samples per cycle from current and voltage waveforms and sends them to a Dynamic Link Library (DLL) block. The general DLL block in PSIM allows users to write codes in C or C++, compile them as a Windows DLL, and link them to PSIM using the features of input (from PSIM) and outputs (returning to PSIM). Unlike the simple DLL blocks with a fixed number of inputs and outputs, the general DLL block provides more flexibility and capability in interfacing PSIM with custom DLL files. In this paper, the DLL codes were used to implement the CPT, the KNN and state-machine NILM, according to the algorithms from Figures 5–7. The state machine from Figure 5 is a loop responsible for the event decision when there is power consumption changing. If there is power changing, the state machine flows in steps until it detects the power event and triggers the "ON event" (Figure 6) or the "OFF event" (Figure 7). If an "ON event" is detected, the system runs algorithm Figure 6 responsible for classifying the appliance. After that, the algorithm adds the appliance to the turned-on appliance list and saves the waveform status for future comparisons when a new event is triggered. If an "OFF event" is detected, the system runs algorithm Figure 7 that is responsible for classifying the appliance and removing the appliance from the turned on appliance list and saves the waveform status for comparison when a new event is triggered.

Simultaneously, the PSB takes care of four other characteristics:

- A database receives the current power consumption in kWh every minute;
- At every turn ON appliance event, the system makes the pattern recognition and sends all the appliance power information to the database, and then, it starts the detected appliance time operation;
- At every turn OFF event, the system recognizes the appliance that was turned off and updates the database, storing the information from the time the load was switched OFF;
- If there are power changes, the system verifies the power signature of each appliance that is turned ON and recognizes it as appliance power level changes.

Figure 8 presents the simulation results. The state machine with the classifier algorithm accepted all the loads according to Table 3. Figure 8 shows the CPT power decomposition and, consequently, the moments in which the algorithm performed the appliance identification (turn on and turn off triggers).

a) Load disaggregation behavior during simulation

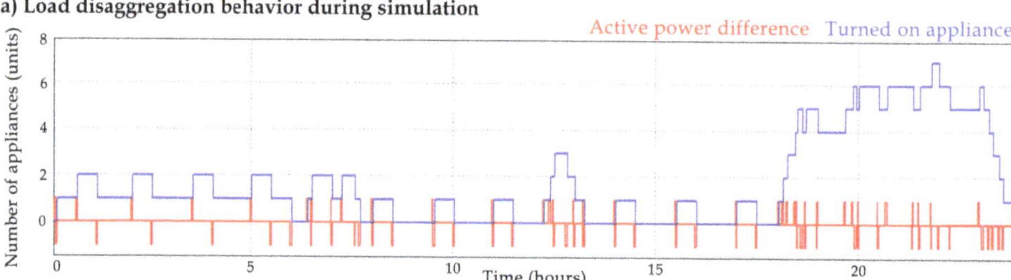

b) CPT power behavior during simulation

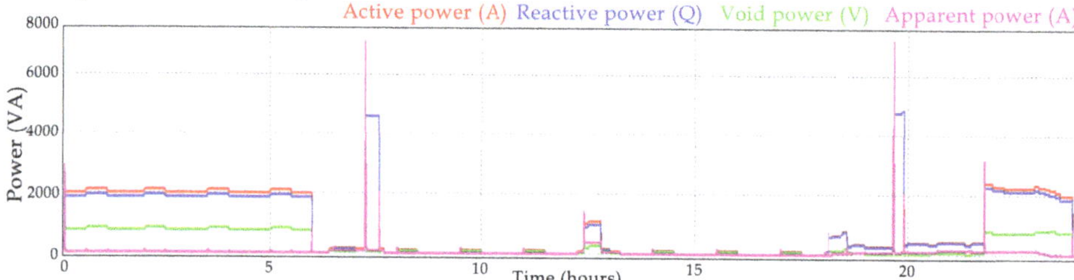

Figure 8. Daily consumption of electricity by household appliances. (**a**) Load disaggregation behavior in simulation; (**b**) CPT power components behavior in simulation.

Figure 9 shows an example of the operation of the PSB between 09:27 and 10:03 from Figure 8. In this interval, there is the operation of a refrigerator, according to the schedule of Table 3. Following the algorithms of Figures 5–7, the state machine has the following behavior:

- **09:27 to 09:30:** Active power is stable, and there is nothing to be done at this time;
- **09:30:** ($\Delta P > Error$); $T_{ON} = TN$, goto 1;
- **09:30:** ($\Delta P > Error$); // waiting for power stabilization;
- **09:31:** ($\Delta P > Error$); // waiting for power stabilization;
- **09:31:** ($\Delta P > Error$); // waiting for power stabilization;
- **09:32:** ($\Delta P < -Error$); goto 5;
- **09:32:** ($\Delta P < Error$) and ($\Delta P > -Error$); ON** Event;

  - CPT Features extraction;
  - KNN classifier;
  - Appliance recognition;
  - Recognized appliance added to the appliance list;
  - goto 0;

- **09:32 to 10:00:** Active power is stable, and there is nothing to be done at this time;
- **10:00:** ($\Delta P < -Error$); $T_{OFF} = TN$; goto 3;
- **10:01:** ($\Delta P < Error$) and ($\Delta P > -Error$); goto 4;
- **10:02:** ($\Delta P < Error$) and ($\Delta P > -Error$); OFF event;

  - Waveform difference;
  - CPT feature extraction;
  - KNN classifier;
  - Appliance recognition;
  - Recognized appliance removed from the appliance list;
  - goto 0;

- **10:02 to 10:03:** Active power is stable, and there is nothing to be done at this time;

a) Load disaggregation behavior between 09:27 to 10:03

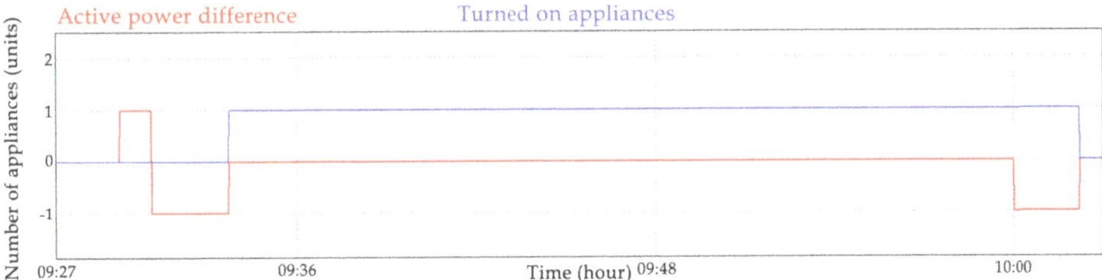

b) Active power behavior between 09:27 to 10:03

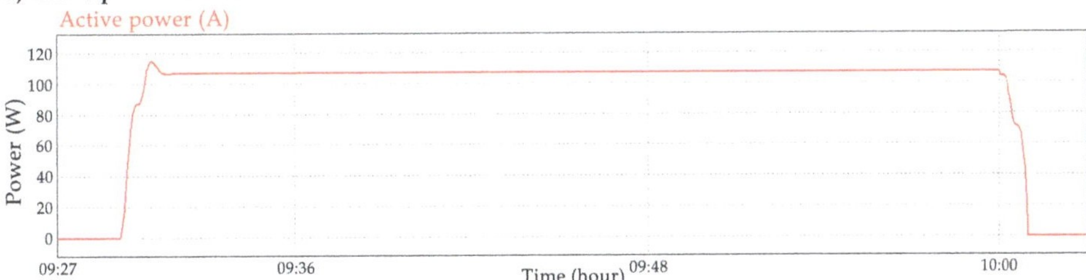

Figure 9. Detailed consumption and state-machine behavior between 09:27 and 10:03. (**a**) Load disaggregation behavior between 09:27 and 10:03; (**b**) Active power behavior between 09:27 and 10:03.

Therefore, the PSB worked as expected, and there was no error in the disaggregation process in the simulations. In this study, there was no error because it is used an appliance dataset with high accuracy and the simulated loads operate in a steady state. However, Section 4.2 will present real appliance cases, including power variations.

4.2. Experimental Results with Household Appliances

The PSB operates with a main loop of 15,360 collected samples. Hence, with a fixed fundamental frequency, this loop takes a second to perform. Then, a four-step procedure tracks rapid power variations. Each step updates the average of the active power evaluated with the last block of samples: 0.25 s of total samples.

Therefore, a predefined threshold level compares the active power variation during the time. Then, the procedure uses the difference between the current average active power and the last one used to define the step level direction: when there is the "power on" or "power off" state of the appliance. Figure 10 shows the behavior of the algorithms from Figures 5–7 in this experiment.

In the trigger events—from Figures 6 and 7—the algorithm stores current and voltage waveforms of the last 0.25 s, extracts the current to be considered in the event and calculates the four elements that represent the appliance: active power, power factor, reactive factor, and non-linearity factor. With these elements (attributes), the algorithm uses the classifier algorithm (the KNN) by means of the knowledge dataset from [15]. The classifier returns the label of the appliance, i.e., the algorithm predicts which appliance is turned on (algorithm from Figure 6) or turned off (algorithm from Figure 7). If it is an "ON event" the recognized appliance is added to the appliance list. If it is an "OFF event", the recognized appliance is removed from the appliance list. After this, the system stores the fifteen waveform cycles of the state to use in a new event trigger. Over time, appliances can be identified, as shown in Figure 11.

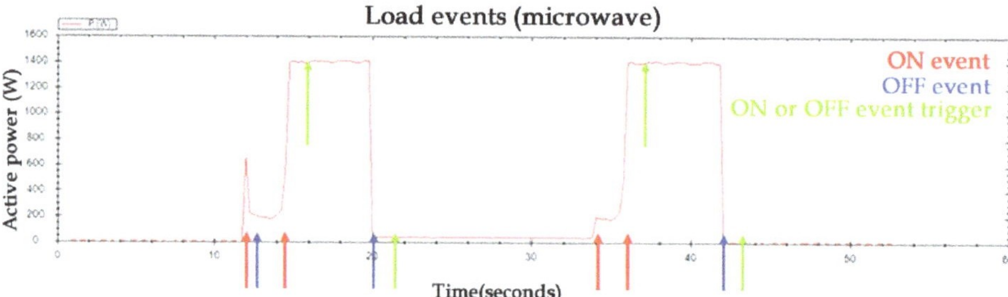

Figure 10. Trigger algorithm validation.

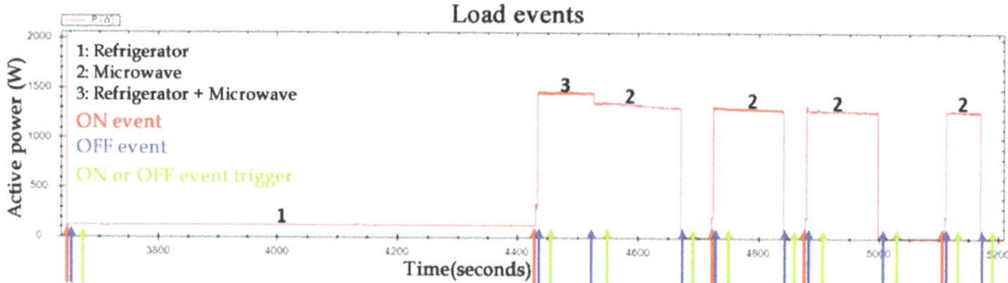

Figure 11. Trigger and load disaggregation validation.

Considering such experiments, the PSB reached 95% of accuracy. The main problem encountered in the method corresponds to the loads with several power changes without the existence of a zero-power instant (i.e., a real turn OFF event). Figure 12 shows an example of an air conditioner with an adjustable speed driver (ASD). In this case, the method of load disaggregation carries out the load identification in the ON event trigger and, in the course of the operation, the power changes without the activation of a new trigger. When the device is turned off, the power level is different from the start of the operation, and the classifier may incorrectly identify the turned off appliance. If there is more than one appliance that is turned on, the algorithm may remove the wrong equipment from the list. If there is only one appliance, the algorithm empties the list.

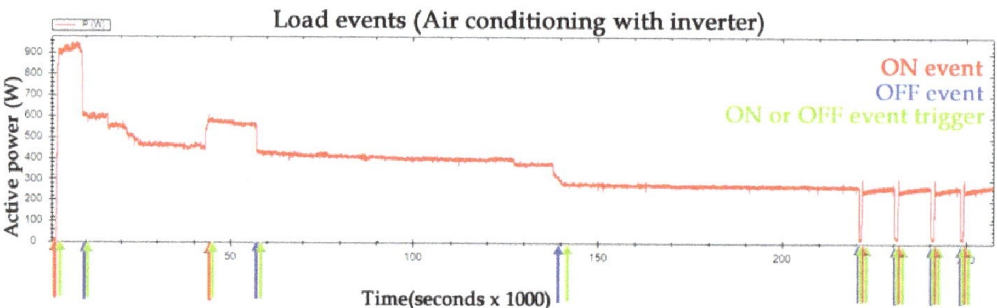

Figure 12. Air conditioning with an inverter example.

To solve this problem, the appliance dataset must have a wide range of appliance measuring times and waveforms should consider the last state (in the OFF event trigger). The algorithm from Figure 5 could also be adapted to improve the accuracy in such situations.

5. Conclusions

This work presented a novel load disaggregation method to be used in cognitive meters, making the "smart" concept from smart meters more reasonable.

The novel methodology called PSB uses the power signature technique with a load recognition from an appliance dataset. The method detects the appliance ON and OFF events during the power signature observation, and the method uses classification algorithms to detect the appliance. The association of the classification algorithm and the power signature recognizes appliances that present power variation during their regular operation. The method also recognizes several loads simultaneously.

When evaluating the proposed method through computational simulations, all the loads were classified correctly. The PSB obtained an accuracy of 95% for real data from a typical residence.

In the future, the authors intend to work on energy efficiency evaluation, associated to identifying appliances. Moreover, the authors would like to apply the PSB to other smart grid applications, such as energy management using the concept of Virtual Power Plants and NILM applied to a group of residential installations. Future papers will deal with such ideas and prominent results.

Author Contributions: Conceptualization, W.A.d.S., F.D.G., F.P.M., L.C.P.d.S. and M.G.S.; Formal analysis, F.P.M.; Funding acquisition, F.P.M. and L.C.P.d.S.; Investigation, W.A.d.S.; Methodology, W.A.d.S., F.D.G. and F.P.M.; Project administration, W.A.d.S. and L.C.P.d.S.; Resources, F.P.M.; Software, W.A.d.S. and F.D.G.; Supervision, F.P.M.; Validation, W.A.d.S., F.D.G. and M.G.S.; Writing—original draft, W.A.d.S.; Writing—review & editing, W.A.d.S., F.D.G., F.P.M., L.C.P.d.S. and M.G.S.

Funding: This research was supported by FAPESP (Sao Paulo Research Foundation) grants number 2012/19375-1 and 2016/08645-9, as well as by the Coordenação de Aperfeiçoamento de Pessoal de Nível Superior—Brasil (CAPES)—Finance Code 001.

Conflicts of Interest: The authors declare no conflict of interest. The funders had no role in the design of the study; in the collection, analyses, or interpretation of data; in the writing of the manuscript, or in the decision to publish the results.

References

1. Tamarkin, T.D. Automatic meter reading. *Public Power* **1992**, *50*, 934–937.
2. Mahmood, A.; Aamir, M.; Anis, M.I. Design and implementation of AMR Smart Grid System. In Proceedings of the IEEE Canada Electric Power Conference, Vancouver, BC, Canada, 6–7 October 2008; pp. 1–6. [CrossRef]
3. Ali, A.; Saad, N.H.; Razali, N.A.; Vitee, N. Implementation of Automatic Meter Reading (AMR) using radio frequency (RF) module. In Proceedings of the IEEE International Conference on Power and Energy, Kota Kinabalu, Malaysia, 2–5 December 2012; pp. 876–879. [CrossRef]
4. Gungor, V.C.; Sahin, D.; Kocak, T.; Ergut, S.; Buccella, C.; Cecati, C.; Hancke, G.P. Smart Grid Technologies: Communication Technologies and Standards. *IEEE Trans. Ind. Inform.* **2011**, *7*, 529–539. [CrossRef]
5. Mohassel, R.R.; Fung, A.; Mohammadi, F.; Raahemifar, K. A survey on Advanced Metering Infrastructure. *Int. J. Electr. Power Energy Syst.* **2014**, *63*, 473–484. [CrossRef]
6. Chien, Y.R.; Yu, H.C. Mitigating Impulsive Noise for Wavelet-OFDM Powerline Communication. *Energies* **2019**, *12*, 1567. [CrossRef]
7. Dong, M.; Meira, P.C.M.; Xu, W.; Chung, C.Y. Non-Intrusive Signature Extraction for Major Residential Loads. *IEEE Trans. Smart Grid* **2013**, *4*, 1421–1430. [CrossRef]
8. Fagiani, M.; Bonfigli, R.; Principi, E.; Squartini, S.; Mandolini, L. A Non-Intrusive Load Monitoring Algorithm Based on Non-Uniform Sampling of Power Data and Deep Neural Networks. *Energies* **2019**, *12*, 1371. [CrossRef]
9. Carroll, J.; Lyons, S.; Denny, E. Reducing household electricity demand through smart metering: The role of improved information about energy saving. *Energy Econ.* **2014**, *45*, 234–243. [CrossRef]
10. Bouhouras, A.S.; Gkaidatzis, P.A.; Chatzisavvas, K.C.; Panagiotou, E.; Poulakis, N.; Christoforidis, G.C. Load Signature Formulation for Non-Intrusive Load Monitoring Based on Current Measurements. *Energies* **2017**, *10*, 538. [CrossRef]

11. · Geelen, D.; Mugge, R.; Silvester, S.; Bulters, A. The use of apps to promote energy saving: A study of smart meter–related feedback in the Netherlands. *Energy Effic.* **2019**. [CrossRef]

12. Chui, K.T.; Lytras, M.D.; Visvizi, A. Energy Sustainability in Smart Cities: Artificial Intelligence, Smart Monitoring, and Optimization of Energy Consumption. *Energies* **2018**, *11*, 2869. [CrossRef]

13. Raj, K.B. Smart Grid Technology for Smart Homes—Risks and Benefits. In Proceedings of the International Conference on Recent Trends in Engineering, Materials, Management and Sciences, Khammam, India, 25–27 October 2018; Swarna Bharathi Institute of Science and Technology: Khammam, India, 2018; Volume 634, pp. 634–640.

14. Hart, G.W. Nonintrusive appliance load monitoring. *Proc. IEEE* **1992**, *80*, 1870–1891. [CrossRef]

15. Souza, W.A.; Marafão, F.P.; Liberado, E.V.; Simões, M.G.; Da Silva, L.C.P. A NILM Dataset for Cognitive Meters Based on Conservative Power Theory and Pattern Recognition Techniques. *J. Control Autom. Electr. Syst.* **2018**, *29*, 742–755. [CrossRef]

16. Zeifman, M. Disaggregation of home energy display data using probabilistic approach. *IEEE Trans. Consum. Electron.* **2012**, *58*, 23–31. [CrossRef]

17. Lin, S.; Zhao, L.; Li, F.; Liu, Q.; Li, D.; Fu, Y. A nonintrusive load identification method for residential applications based on quadratic programming. *Electr. Power Syst. Res.* **2016**, *133*, 241–248. [CrossRef]

18. Du, L.; He, D.; Harley, R.G.; Habetler, T.G. Electric Load Classification by Binary Voltage–Current Trajectory Mapping. *IEEE Trans. Smart Grid* **2016**, *7*, 358–365. [CrossRef]

19. Gillis, J.M.; Alshareef, S.M.; Morsi, W.G. Nonintrusive Load Monitoring Using Wavelet Design and Machine Learning. *IEEE Trans. Smart Grid* **2016**, *7*, 320–328. [CrossRef]

20. Le, T.T.H.; Kim, H. Non-Intrusive Load Monitoring Based on Novel Transient Signal in Household Appliances with Low Sampling Rate. *Energies* **2018**, *11*, 3409. [CrossRef]

21. He, H.; Liu, Z.; Jiao, R.; Yan, G. A Novel Nonintrusive Load Monitoring Approach based on Linear-Chain Conditional Random Fields. *Energies* **2019**, *12*, 1797. [CrossRef]

22. Kwak, Y.; Hwang, J.; Lee, T. Load Disaggregation via Pattern Recognition: A Feasibility Study of a Novel Method in Residential Building. *Energies* **2018**, *11*, 1008. [CrossRef]

23. Tenti, P.; Mattavelli, P.; Morales Paredes, H.K. Conservative Power Theory, sequence components and accountability in smart grids. In Proceedings of the International School on Nonsinusoidal Currents and Compensation, Lagow, Poland, 15–18 June 2010; pp. 37–45. [CrossRef]

24. Tenti, P.; Paredes, H.K.M.; Mattavelli, P. Conservative Power Theory, a Framework to Approach Control and Accountability Issues in Smart Microgrids. *IEEE Trans. Power Electron.* **2011**, *26*, 664–673. [CrossRef]

25. Souza, W.A.; Liberado, E.V.; da Silva, L.C.P.; Paredes, H.K.M.; Marafão, F.P. Load analyser using conservative power theory. In Proceedings of the International School on Nonsinusoidal Currents and Compensation, Zielona Gora, Poland, 20–21 June 2013; pp. 1–6. [CrossRef]

26. Drenker, S.; Kader, A. Nonintrusive monitoring of electric loads. *IEEE Comput. Appl. Power* **1999**, *12*, 47–51. [CrossRef]

27. Cole, A.I.; Albicki, A. Data extraction for effective non-intrusive identification of residential power loads. In Proceedings of the IEEE Instrumentation and Measurement Technology Conference, St. Paul, MN, USA, 18–21 May 1998; Volume 2, pp. 812–815.

28. Leeb, L.K.N.S.B. Non-intrusive electrical load monitoring in commercial buildings based on steady-state and transient load-detection algorithms. *Energy Build.* **1996**, *24*, 51–64. [CrossRef]

29. Plungklang, S.B.B. Non-Intrusive Appliances Load Monitoring (NILM) for Energy Conservation in Household with Low Sampling Rate. *Procedia Comput. Sci.* **2016**, *86*, 172–175. [CrossRef]

30. Xiao, P.; Cheng, S. Neural Network for NILM Based on Operational State Change Classification. *arXiv* **2019**, arXiv:1902.02675.

31. Powers, J.T.; Margossian, B.; Smith, B.A. Using a rule-based algorithm to disaggregate end-use load profiles from premise-level data. *IEEE Comput. Appl. Power* **1991**, *4*, 42–47. [CrossRef]

32. Zmeureanu, L.F.R. Using a pattern recognition approach to disaggregate the total electricity consumption in a house into the major end-uses. *Energy Build.* **1999**, *30*, 245–259. [CrossRef]

33. Zmeureanu, M.M.R. Nonintrusive load disaggregation computer program to estimate the energy consumption of major end uses in residential buildings. *Energy Convers. Manag.* **2000**, *41*, 1389–1403. [CrossRef]

34. Baranski, M.; Voss, J. Genetic algorithm for pattern detection in NIALM systems. In Proceedings of the IEEE International Conference on Systems, Man and Cybernetics, The Hague, The Netherlands, 10–13 October 2004; Volume 4, pp. 3462–3468. [CrossRef]

35. Ruzzelli, A.G.; Nicolas, C.; Schoofs, A.; O'Hare, G.M.P. Real-Time Recognition and Profiling of Appliances through a Single Electricity Sensor. In Proceedings of the IEEE Communications Society Conference on Sensor, Mesh and Ad Hoc Communications and Networks, Boston, MA, USA, 21–25 June 2010; pp. 1–9. [CrossRef]

36. Kelly, J.; Knottenbelt, W. Neural NILM: Deep Neural Networks Applied to Energy Disaggregation. In Proceedings of the ACM International Conference on Embedded Systems for Energy-Efficient Built Environments, Seoul, Korea, 4–5 November 2015; ACM: New York, NY, USA, 2015; pp. 55–64. [CrossRef]

37. Figueiredo, M.B.; de Almeida, A.; Ribeiro, B. An Experimental Study on Electrical Signature Identification of Non-Intrusive Load Monitoring (NILM) Systems. In *Adaptive and Natural Computing Algorithms*; Springer: Berlin/Heidelberg, Germany, 2011; pp. 31–40.

38. Kim, H.; Marwah, M.; Arlitt, M.; Lyon, G.; Han, J. Unsupervised Disaggregation of Low Frequency Power Measurements. In Proceedings of the SIAM International Conference on Data Mining, Mesa, AZ, USA, 28–30 April 2011; pp. 747–758. [CrossRef]

39. Koutitas, G.C.; Tassiulas, L. Low Cost Disaggregation of Smart Meter Sensor Data. *IEEE Sens. J.* **2016**, *16*, 1665–1673. [CrossRef]

40. Sultanem, F. Using appliance signatures for monitoring residential loads at meter panel level. *IEEE Trans. Power Deliv.* **1991**, *6*, 1380–1385. [CrossRef]

41. Srinivasan, D.; Ng, W.S.; Liew, A.C. Neural-network-based signature recognition for harmonic source identification. *IEEE Trans. Power Deliv.* **2006**, *21*, 398–405. [CrossRef]

42. Laughman, C.; Lee, K.; Cox, R.; Shaw, S.; Leeb, S.; Norford, L.; Armstrong, P. Power signature analysis. *IEEE Power Energy Mag.* **2003**, *99*, 56–63. [CrossRef]

43. Bouhouras, A.S.; Gkaidatzis, P.A.; Panagiotou, E.; Poulakis, N.; Christoforidis, G.C. A NILM algorithm with enhanced disaggregation scheme under harmonic current vectors. *Energy Build.* **2019**, *183*, 392–407. [CrossRef]

44. Teshome, D.F.; Huang, T.D.; Lian, K. Distinctive Load Feature Extraction Based on Fryze's Time-Domain Power Theory. *IEEE Power Energy Technol. Syst. J.* **2016**, *3*, 60–70. [CrossRef]

45. Nguyen, M.; Alshareef, S.; Gilani, A.; Morsi, W.G. A novel feature extraction and classification algorithm based on power components using single-point monitoring for NILM. In Proceedings of the IEEE Canadian Conference on Electrical and Computer Engineering, Halifax, NS, Canada, 3–6 May 2015; pp. 37–40. [CrossRef]

46. Huang, T.D.; Wang, W.; Lian, K. A New Power Signature for Nonintrusive Appliance Load Monitoring. *IEEE Trans. Smart Grid* **2015**, *6*, 1994–1995. [CrossRef]

47. Chan, W.L.; So, A.T.P.; Lai, L.L. Harmonics load signature recognition by wavelets transforms. In Proceedings of the International Conference on Electric Utility Deregulation and Restructuring and Power Technologies, London, UK, 4–7 April 2000; pp. 666–671. [CrossRef]

48. Su, Y.; Lian, K.; Chang, H. Feature Selection of Non-intrusive Load Monitoring System Using STFT and Wavelet Transform. In Proceedings of the IEEE International Conference on e-Business Engineering, Beijing, China, 19–21 October 2011; pp. 293–298. [CrossRef]

49. Duarte, C.; Delmar, P.; Goossen, K.W.; Barner, K.; Gomez-Luna, E. Non-intrusive load monitoring based on switching voltage transients and wavelet transforms. In Proceedings of the Future of Instrumentation International Workshop, Gatlinburg, TN, USA, 8–9 October 2012; pp. 1–4. [CrossRef]

50. Gray, M.; Morsi, W.G. Application of wavelet-based classification in non-intrusive load monitoring. In Proceedings of the IEEE Canadian Conference on Electrical and Computer Engineering, Halifax, NS, Canada, 3–6 May 2015; pp. 41–45. [CrossRef]

51. Tabatabaei, S.M.; Dick, S.; Xu, W. Toward Non-Intrusive Load Monitoring via Multi-Label Classification. *IEEE Trans. Smart Grid* **2017**, *8*, 26–40. [CrossRef]

52. Chang, H.H. Non-Intrusive Demand Monitoring and Load Identification for Energy Management Systems Based on Transient Feature Analyses. *Energies* **2012**, *5*, 4569–4589. [CrossRef]

53. Chang, H.; Lian, K.; Su, Y.; Lee, W. Power-Spectrum-Based Wavelet Transform for Nonintrusive Demand Monitoring and Load Identification. *IEEE Trans. Ind. Appl.* **2014**, *50*, 2081–2089. [CrossRef]

54. Hassan, T.; Javed, F.; Arshad, N. An Empirical Investigation of V-I Trajectory Based Load Signatures for Non-Intrusive Load Monitoring. *IEEE Trans. Smart Grid* **2014**, *5*, 870–878. [CrossRef]
55. Gao, J.; Kara, E.C.; Giri, S.; Bergés, M. A feasibility study of automated plug-load identification from high-frequency measurements. In Proceedings of the IEEE Global Conference on Signal and Information Processing, Orlando, FL, USA, 14–16 December 2015; pp. 220–224. [CrossRef]
56. De Baets, L.; Ruyssinck, J.; Develder, C.; Dhaene, T.; Deschrijver, D. Appliance classification using VI trajectories and convolutional neural networks. *Energy Build.* **2018**, *158*, 32–36. [CrossRef]
57. Patel, S.N.; Robertson, T.; Kientz, J.A.; Reynolds, M.S.; Abowd, G.D. *At the Flick of a Switch: Detecting and Classifying Unique Electrical Events on the Residential Power Line (Nominated for the Best Paper Award);* UbiComp 2007 Ubiquitous Computing; Springer: Berlin/Heidelberg, Germany, 2007; pp. 271–288.
58. Gupta, S.; Reynolds, M.S.; Patel, S.N. ElectriSense: Single-point Sensing Using EMI for Electrical Event Detection and Classification in the Home. In Proceedings of the ACM International Conference on Ubiquitous Computing, Copenhagen, Denmark, 26–29 September 2010; pp. 139–148. [CrossRef]
59. Kolter, J.Z.; Johnson, M.J. REDD: A public data set for energy disaggregation research. In Proceedings of the Workshop on Data Mining Applications in Sustainability, San Diego, CA, USA, 21 August 2011; Volume 25, pp. 59–62.
60. Kelly, J.; Knottenbelt, W. The UK-DALE dataset, domestic appliance-level electricity demand and whole-house demand from five UK homes. *Sci. Data* **2015**, *2*, 150007. [CrossRef]
61. Gao, J.; Giri, S.; Kara, E.C.; Bergés, M. PLAID: A Public Dataset of High-resoultion Electrical Appliance Measurements for Load Identification Research: Demo Abstract. In Proceedings of the ACM Conference on Embedded Systems for Energy-Efficient Buildings, Memphis, Tennessee, 3–6 November 2014; pp. 198–199. [CrossRef]
62. Kahl, M.; Haq, A.U.; Kriechbaumer, T.; Jacobsen, H.A. Whited-a worldwide household and industry transient energy data set. In Proceedings of the International Workshop on Non-Intrusive Load Monitoring, Vancouver, BC, Canada 14–15 May 2016.
63. Papa, J.P.; Falcão, A.X.; Suzuki, C.T.N. Supervised pattern classification based on optimum-path forest. *Int. J. Imaging Syst. Technol.* **2009**, *19*, 120–131. [CrossRef]
64. Cover, T.M.; Hart, P.E. Nearest neighbor pattern classification. *IEEE Trans. Inf. Theory* **1967**, *13*, 21–27. [CrossRef]
65. Tenti, P.; Paredes, H.K.M.; Marafao, F.P.; Mattavelli, P. Accountability in Smart Microgrids Based on Conservative Power Theory. *IEEE Trans. Instrum. Meas.* **2011**, *60*, 3058–3069. [CrossRef]
66. Bouhouras, A.S.; Milioudis, A.N.; Labridis, D.P. Development of distinct load signatures for higher efficiency of NILM algorithms. *Electr. Power Syst. Res.* **2014**, *117*, 163–171. [CrossRef]

Article

Knowledge Embedded Semi-Supervised Deep Learning for Detecting Non-Technical Losses in the Smart Grid

Xiaoquan Lu [1,2], Yu Zhou [1,2], Zhongdong Wang [1,2], Yongxian Yi [1,2], Longji Feng [3] and Fei Wang [4,*

[1] State Grid Jiangsu Electric Power Co., Ltd. Research Institute, Nanjing 210019, China; seueelab_lxq@163.com (X.L.); jsepc_zy@163.com (Y.Z.); llpppxmtd@163.com (Z.W.); wzcnyyx@163.com (Y.Y.)

[2] State Grid Key laboratory of Electrical Power Metering, Nanjing 210039, China

[3] State Grid Nanjing Power Supply Company, Nanjing 210000, China; flj95598@163.com

[4] School of Electronic Information and Communications, Huazhong University of Science and Technology, Wuhan 430074, China

* Correspondence: wangfei@hust.edu.cn

Received: 23 July 2019; Accepted: 5 September 2019; Published: 6 September 2019

Abstract: Non-technical losses (NTL) caused by fault or electricity theft is greatly harmful to the power grid. Industrial customers consume most of the power energy, and it is important to reduce this part of NTL. Currently, most work concentrates on analyzing characteristic of electricity consumption to detect NTL among residential customers. However, the related feature models cannot be adapted to industrial customers because they do not have a fixed electricity consumption pattern. Therefore, this paper starts from the principle of electricity measurement, and proposes a deep learning-based method to extract advanced features from massive smart meter data rather than artificial features. Firstly, we organize electricity magnitudes as one-dimensional sample data and embed the knowledge of electricity measurement in channels. Then, this paper proposes a semi-supervised deep learning model which uses a large number of unlabeled data and adversarial module to avoid overfitting. The experiment results show that our approach can achieve satisfactory performance even when trained by very small samples. Compared with the state-of-the-art methods, our method has achieved obvious improvement in all metrics.

Keywords: non-technical losses; smart grid; semi-supervised learning; knowledge embed; deep learning

---

1. Introduction

Non-technical losses (NTL) are one of the most major problems pertaining to the power grid, and have been for quite a long time. Unlike technical losses which are generally caused during generation and distribution, NTL are anomalies which include installation errors, faulty meters and electricity theft, etc. Referring to World Bank reports, NTL represents a significant part of the total power losses in both developing and developed nations [1]. A survey from the Northeast Group LLC shows that more than \$89.3 billion is lost every year worldwide due to NTL [2]. Besides financial losses, NTL also causes a decrease of stability and reliability of the power grid.

Presently, over 80% of the global population has access to electricity [1]. However, in total electricity consumption, industrial and large commercial customers contribute approximately 55% in Spain [3]. Similarly in China, the ratio of industrial customers is more than 65% [4]. Naturally, detecting NTL among industrial customers is more interesting than residential customers to electricity providers. Hence, this paper aims to detect NTL among industrial customers.

Conventional NTL detection methods depend on the in-field inspection, where both the costs and efficiency can not satisfy electricity providers. With the appearance of the smart grid comes a great deal of smart meter (SM) data and extra opportunities to solve NTL. Hence, a lot of data oriented methods have been proposed recently, due to the development of machine learning and ease of implementation [5]. Researchers adopt methods of different fields of knowledge with machine learning, such as anomaly detection, cybersecurity, etc. Generally, these approaches can be classified as supervised, unsupervised and ensemble methods. Through studying anomaly behaviour in electricity consumption, some of them can help to identify NTL indeed [6]. However, they only got better effect on residential customers rather than industrial customers. The primary reasons are listed as follows:

1. **The consumption pattern of residential customers is more stable than that of industrial customers**. It is easier to find change points of residential consumption history, while industrial customers have multiple consumption patterns because they have to adjust their consumption behaviour according to the market [7].
2. **Residential customers are similar to each other, while industrial customers are quite different**. It is easier to cluster residential customers into limited categories [8]. Particularly, [9] uses the location as assistant feature. On the contrary, industrial customers' consumption patterns are quite different from one another, even if they belong to the same domain or are located near to each other [3].

Therefore, it is more difficult to detect NTL only depends on electricity consumption among industrial customers. The key challenges are reflected on the follows:

1. **How to extract features with higher linear separability?** Refer to recent research, the features are mostly designed manually according to observation and experience based on electricity consumption. They can hardly represent all scenarios of NTL because consumption behaviour is random and unpredictable, especially for industrial customers. For example, when changes in consumption pattern due to change in household residents or usage of electrical devices might make it looks like electricity theft [10]. This situation would destroy linear separability of traditional features.
2. **How to obtain satisfactory performance based on limited labeled samples?** Compared to unsupervised learning-based methods, supervised learning-based methods acquired better detection accuracy and become used in the mainstream gradually. However, the realistic NTL samples of which in-field inspected are rare indeed, supervised learning methods are an easier lead to overfitting. On the other hand, the artificial samples as a possible solution are adopted by some approaches [10,11]. Even though they provide lots of labeled samples to support training models, the effectiveness of such attack models is not verified by realistic cases. Due to preconfigured parameters or fixed distribution, it is believable that such artificial samples would also lead to overfitting easily.
3. **How to achieve higher accuracy among various customers?** Currently, many of related approaches which are based on Support Vector Machines (SVM), K-Nearest Neighbors (KNN), etc., have low accuracy. Even if using much auxiliary data [3] or a large number of labeled samples [12], the performance of these approaches still cannot suit realistic requirements.

Therefore, this paper proposes a deep learning-based Semi-Supervised AutoEncoder (SSAE) model, and attempts to solve the above problems and achieve an ideal NTL recognition accuracy. In this work, we focus on three-phase industrial customers with a contracted power higher than 80 kVA. We design a deep semi-supervised neural network to learn advanced features from massive SM data includes voltage, current, active power, etc. The extracted features cover both principle of electricity measurement and consumption behavior through knowledge embedding. Our model has been trained, validated and tested using real in-field inspected data. Overall, the main contributions of this paper can be summarized as follows:

1. Based on SM data, this paper designs a domain knowledge embedded data model to enhance linear separability of normal samples and abnormal samples.

2. We propose a novel deep neural network-based semi-supervised model to extract advanced features from limited labeled samples efficiently. In addition, by designing an adversarial module, our model has stronger anti-noise ability.
3. Our approach improves the performance of NTL detection obviously. Experimental results show that all metrics of SSAE model outperform other existing approaches in realistic cases.

The remainder of the paper is organized as follows. Section 2 presents a brief overview of NTL detection. Section 3 presents the problem analysis and introduces knowledge embedded data model. Section 4 presents deep semi-supervised model. Experiments are conducted and the evaluation results are shown in Section 5. Finally, we provide conclusions and future work in Section 6.

2. Related Work

The recent research for NTL detection is around hardware or non-hardware solutions. Due to hardware based solution needing further special sensors, the cost and efficiency can hardly satisfy electricity providers even if it has higher accuracy [12]. With the growing of the smart grid and implementation of advanced metering infrastructure (AMI) systems, electricity providers collect and hold various and massive SM data. Hence, non-hardware solutions are more acceptable to electricity providers, especially data oriented methods. Hence, this section only presents a brief survey of the state-of-the-art on data oriented NTL detection methods.

According to implemented machine learning algorithms, data oriented methods could be roughly categorized into three types:

1. **Supervised learning-based methods**. They mainly include Decision Tree (DT) [13,14], Support Vector Machines (SVM) [13,15,16], K-Nearest Neighbors (KNN) [17], Bayesian Networks (BN) [18], Artificial Neural Network(ANN) [12,17], Deep Neural Network(DNN) [12], etc. Depend on supervised learning algorithm and artificial feature, these methods acquire satisfactory effect in some situations. Due to the variety of NTL, especially electricity theft, these methods require large samples with the right labels to train algorithms. However, it is difficult to collect enough realistic normal and abnormal cases, which makes the labeled samples are very scarce. To avoid insufficient realistic NTL samples, [10,11,13] attempt to model NTL and produce artificial NTL samples. Same as the necessary requirement of massive labeled samples, features are equally important to classifiers. To construct powerful feature models, most researchers applied raw data, statistics, Fourier coefficients, wavelet coefficients, slope of consumption curve, etc. However, all of them still could not cover all situation of NTL, especially the NTL of industrial customers. Hence, [12] proposed deep learning-based method and self-learned features from massive consumption data. The results from [12] show that wide and deep convolutional neural networks has strong feature learning ability and improve accuracy in electricity theft detection. However, the limitation of consumption data blocks the implementation of [12] to industrial customers.
2. **Unsupervised learning-based methods**. To avoid labeling massive samples, some researchers choose unsupervised methods to detect NTL. Unsupervised methods do not need any labeled samples, and primary contain clustering [8,19], outlier detection [20] and expert systems [21]. Even though a series of unsupervised methods are free from labeling training set, their performance always could hardly satisfy electricity providers when they deploy standalone. Frequently, unsupervised algorithms pose auxiliary methods. They are used to group similar consumers, and then train further classifiers on these groups [8,19]. However, because of the fact that abnormal samples are always far less than the normal samples, it is still difficult to promise that each group own enough labeled samples. Hence, misjudging tends to occur whatever clustering or classification.
3. **Semi-supervised learning-based methods**. They allow the NTL detector to be trained on a few labeled samples and large unlabeled samples [22]. In ref. [23] uses Transductive SVM(TSVM) to build a NTL detecting system. Restricted to TSVM could not handle imbalance situation, [23] has

not been demonstrated enough for detecting NTL. However, semi-supervised learning still is competitive and hopeful choice for detecting NTL when it meets deep learning.

Summary, most proposals meet following limitations: (1) Electricity consumption is not enough to classify normal and abnormal cases in all possible scenarios; (2) Artificial NTL samples are different from realistic cases and lose effect on industrial customers; (3) Performance of these methods still needs to be greatly improved.

In the recent years, the field of machine-learning has produced several pivotal advances that address complex problems. Deep learning simulates the brain's structure with multiple layers of neurons, fitting complex functions, and characterizing the input data's distribution, has demonstrated excellent capacity of automatically learning features. It is widely adopted in computer vision [24], speech-recognition [25], natural language processing [26], etc. and has achieved huge success. Simultaneously, a sequence of semi-supervised deep learning models [27–29] have been proposed. It is demonstrated that they had achieved remarkable success in image classification tasks. So, great potential exists that deep learning would contribute a lot to NTL detection application, the research about which is just in the beginning phase.

The electrical magnitudes are great different from data which they handled. Firstly, electrical magnitudes are consisting of multiple time series data, such as voltage, current, etc. Secondary, the dimension of electrical magnitudes is significantly different from image and audio. Furthermore, the knowledge has been naturally embedded in the picture or audio, however, electrical magnitudes do not contain any domain knowledge. Therefore, this study attempts to propose a novel semi-supervised deep learning model to overcome the limitations of the above existing works and detect NTL accurately.

3. Modeling Samples with Knowledge

Although deep learning has strong feature learning ability, it is still difficult to learn domain knowledge from raw data. Hence, this section will provide an efficient way to embed knowledge of electricity measurement into sample model to help deep neural network to extract advanced features.

3.1. Principle of Electricity Measurement

Most industrial customers equipped with three-phase smart meters. According to the different wiring modes, it can be divided into two types: three-phase four-wire and three-phase three-wire. In this paper, we primary introduce our approach with three-phase four-wire as an example. For single phase scenario, the electricity energy is calculated by following equation:

$$E = \sum_{n=0}^{N-1} P_n \cdot \Delta t \tag{1}$$

where, Δt is the cycle of computation, E is the active electricity energy in a certain time period $N \cdot \Delta t$, P_n is the average active power at time n. It can be further calculated by voltage and current:

$$P_n = U_n \cdot I_n \cdot cos\phi_n \tag{2}$$

where, U_n and I_n are average voltage and current at time n, ϕ_n is the phase difference between voltage and current at the same time stamp and $cos\phi_n$ is called power factor. Further, the equation of active energy are rewritten by:

$$E = \sum_{n=0}^{N-1} [U_n \cdot I_n \cdot cos\phi_n] \cdot \Delta t \tag{3}$$

For three-phase four-wire situation, the total active electricity energy is calculated by:

$$E_{total} = \sum_{n=0}^{N-1} P_{total} \cdot \Delta t = \sum_{n=0}^{N-1} [P_A^n + P_B^n + P_C^n] \cdot \Delta t \tag{4}$$

Generally, electricity energy measured by voltage and current, and the relationship between them is very important to NTL detection.

3.2. SM Data

Due to industrial customers contribute most electricity consumption, electrical providers equipped AMI system to read, collect and save important electricity magnitudes. These magnitudes are read every 15 min and all of them share the same time-stamp. Based on Equations (1)–(3), this paper collects part of them as dataset to train, validate and test the model from State Grid Corporation of China(SGCC) which listed in Table 1. The SM data are labeled manually refer to the result of in-field inspection. The rule for data labeling is that all data of each customer has the same label. There are few label noise in our data set because abnormal customers were not always abnormal.

Table 1. Electrical magnitudes of three-phase four-wire.

Magnitude	Description
U_A, U_B, U_C	average voltage of each wire
I_A, I_B, I_C	average current of each wire
P_{total}	total active power
Q_{total}	total reactive power
f_{total}	total power factor
ts	time-stamp

3.3. Analysis of NTL

The collected SM data reflects the electricity consumption at a certain moment. For normal situation, SM data must follow the principle of electricity measurement mentioned above. On the contrary, anomaly SM data must break the regular law to reduce consumption randomly. Refer to the principle of electricity measurement, the primary types of NTL include:

1. **Shunts**: For a three-phase situation, the shunts is the outputs of PT or CT are shorted or injected a low-resistance path. In general, it would reduce the measured voltage or current. Commonly, voltages and currents of three-phase customers are almost balance [30], such as Figure 1a,c. The shunts will break the balance on voltages or currents, such as Figure 1b,d. In particular, the degree of imbalance about three-phase currents is related to customers' load level, and is different among all customers. Shunts are the most complicated NTL because normal customers may also have similar phenomenon. In Figure 1c, the currents of a normal customer also have little imbalance when the load is low.
2. **Phase Shift**: It means that the phase difference between voltage and current is changed artificially. Through increasing ϕ to reduce the power factor, and decrease the measured active power. There is a significant phenomenon that the power factor reduced obviously. Commonly, the power factor should close to 1.
3. **Phase disorder**: The currents are coupled with the wrong voltages. Figure 1e–h presents a typical example of this case. The output of phase-A's CT and PT are jointed mistakenly with the output of phase-B's PT and CT. From the curves of Figure 1e–g, it can be found that the voltages and currents and power factor are almost normal respectively. However, there is a large gap between measured total active power and estimated total apparent power in Figure 1h.
4. **Phase Inversion**: The phase of voltage or current is inverse directly. The phase difference between voltage and current is changed from ϕ to $(\pi - \phi)$. The smart meter will measure a negative active power of such wire and smaller total active power. Figure 2 shows a more complicated situation, all phases are inversion. This time, there is no abnormality in SM data, too. However, when we analyze active power and power factor jointly, it can be found that they are negative related rather than positive related.

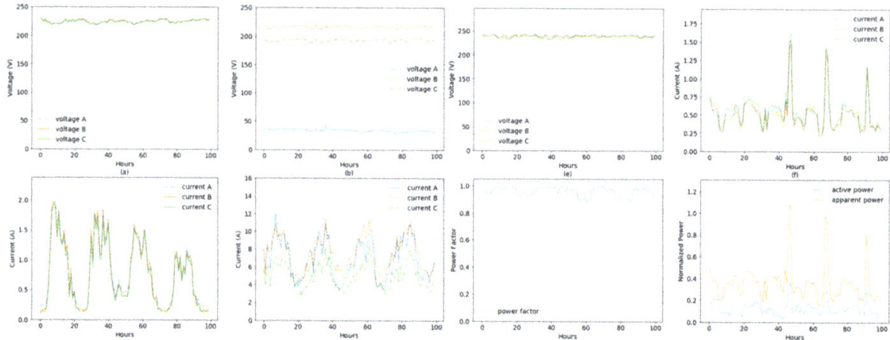

Figure 1. Comparision of normal and Shunts by curves of voltages and currents.

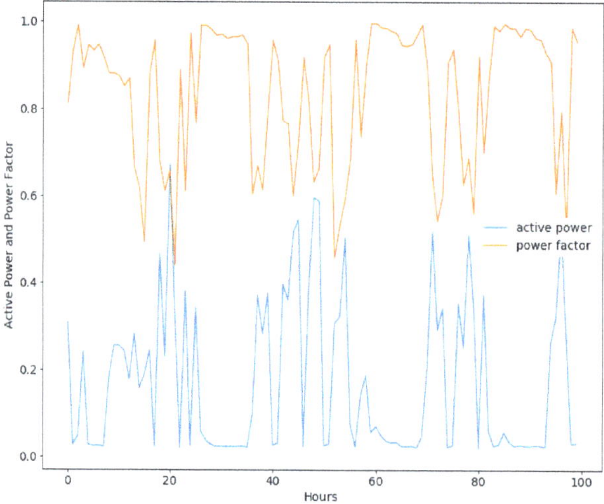

Figure 2. Relationship between active power and power factor about phase inversion.

Overall, it is difficult to discover NTL among industrial customers only with SM data because the pattern of NTL mentioned above will be changed timely and randomly. Furthermore, it might be existed from the beginning, such as phase disorder. In contrast to the SMART attack defined in [11] or FDI5 defined in [10], the realistic NTL does not run in fixed artificial model and is more random and complicated. Consequently, the domain knowledge based features are very important for classifier to detect NTL.

3.4. Sample Model with Knowledge Embedding

Before starting to train a deep neural network (DNN), it is necessary to model SM data as a suitable format. In [12], the electricity consumption is organized as a 1-D vector or 2-D matrix to feed DNN. It is different in that the SM data are multiple time series data. This paper tends to organize them as a vector with multiple channels. After comparing varying time span, we found weekly SM data has better performance. Hence, we choose week as the time span of sample and design a shifting window to construct different samples which shown in Figure 3a. The samples within same customer own the same label.

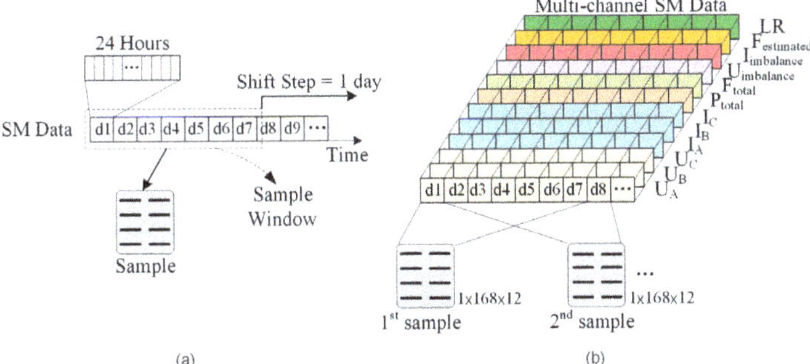

Figure 3. Structure of sample. (**a**) create samples based on shift window. (**b**) sample model embedded by knowledge.

In Section 3.3, we analyzed the complexity of NTL and came to the preliminary conclusion that only SM data is not enough to detect it. To further evaluating the linear separability of the samples based on pure SM data, we use t-SNE algorithm [31] to visualize samples and the result is shown in Figure 4a. There are lots of abnormal samples spread in the range of normal samples. They must lead to worse performance of NTL detection. To improve the linear separability of the samples, this paper attempts to embed electrical knowledge into the sample model. Based on principle of electricity measurement and phenomenon of NTL, we use the following parameters as additional channels:

$$U_{imbalance} = \frac{max(U_A, U_B, U_C) - min(U_A, U_B, U_C)}{max(U_A, U_B, U_C)} \tag{5}$$

$$I_{imbalance} = \frac{max(I_A, I_B, I_C) - min(I_A, I_B, I_C)}{max(I_A, I_B, I_C)} \tag{6}$$

$$\hat{f} = \frac{P_{total}}{U_A \cdot I_A + U_B \cdot I_B + U_C \cdot I_C} \tag{7}$$

$$LR = \frac{U_A \cdot I_A + U_B \cdot I_B + U_C \cdot I_C}{Contracted\ Apparent\ Power} \tag{8}$$

Finally, the sample is organized as multi-channel vector which is shown in Figure 3b. From Figure 4b, it is easy to find that the linear separability of the samples embedded knowledge is improved obviously, only few abnormal samples overlap with the normal samples.

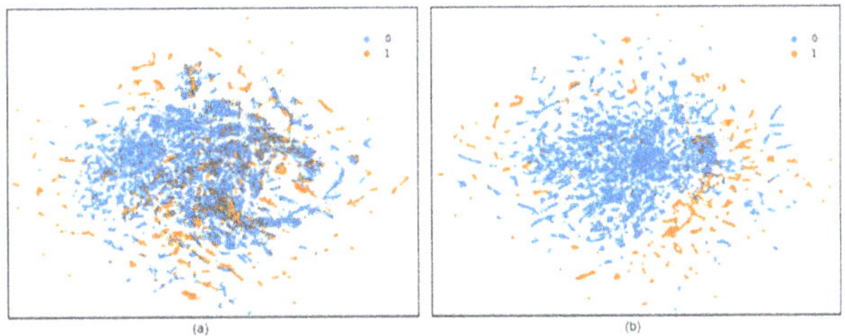

Figure 4. t-SNE results of samples where 0 denotes normal samples plotted in blue, and 1 denotes NTL samples plotted in orange. (**a**) samples based on raw SM data. (**b**) samples based on raw SM data and electricity measurement knowledge.

3.5. Data Preprocess

Because there are some missing and error values in SM data caused by communication error or smart meter failure, the raw data cannot be used by DNN directly. For missing data, this paper interpolates them by the average of before and after 2 days at the same hour of day. The detailed equation is shown by:

$$\hat{x}_i = \frac{\sum_{k=-2}^{2} x_{i+(24*k)}}{\#(Existed)} \tag{9}$$

where, $\hat{x}_i$ is the interpolated value, i denotes the hourly time stamp, $\#(Existed)$ represents the number of existed values in before and after two days at the same hour of day.

For error values, there are two situations: (1) negative value; (2) extreme value. We process them by following equation:

$$x_i = \begin{cases} 0 & x_i < 0 \\ Q_3(x) + [Q_3(x) - Q_1(x)] * 3 & x_i > Q_3(x) + [Q_3(x) - Q_1(x)] * 3 \end{cases} \tag{10}$$

where, x is a sequence of certain electricity magnitude of certain customer, $Q_3(\cdot)$ and $Q_1(\cdot)$ are upper quantile and lower quantile respectively. For different electricity magnitude or different customer, $Q_3(\cdot)$ and $Q_1(\cdot)$ are different.

Furthermore, standardization of samples is a necessary requirement for most machine learning algorithms, especially DNN. Too large value of samples will cause excessive computing error for DNN. According to SM data, different magnitudes have different scales, especially among different customers. Non-uniform scalers of samples will degrade the predictive performance of DNN. Refer to Sections 3.1 and 3.3, the values of f_{total}, $U_{imbalance}$, $I_{imbalance}$, $\hat{f}$, LR are already located in the range of $[0, 1]$. Hence, the remainder voltages and currents are normalized respectively according to following equation:

$$\bar{x}_i = \frac{x_i}{Q_3(x) + [Q_3(x) - Q_1(x)] * 3} \tag{11}$$

where, $\bar{x}_i$ is normalized value, x is a sequence of certain electricity magnitude of certain customer. After this, all channels of a sample and all samples from different customers are normalized into same range.

4. Semi-Supervised AutoEncoder

Besides knowledge embedded sample model, a powerful semi-supervised deep neural network is also necessary to extract advanced features from limited labeled samples. In this section, we will introduce the framework and algorithm of the Semi-Supervised AutoEncode r(SSAE) to show how it works.

4.1. Framework of SSAE

The SSAE is a generative model. It consists of four modules: encoder, decoder, discriminator and classifier. In this model, the encoder and decoder are coupled as an autoencoder. They could learn more general features from all samples include unlabeled and labeled. Due to generative model, the SSAE could avoid overfitting effectively. The discriminator is aimed to regularize the autoencoder by a specified arbitrary prior. It judges the encoding distribution of X is same as the prior or not. This idea is borrowed from [27]. The classifier are designed to select features from latent vector and classify normal and abnormal. The architecture of SSAE is shown in Figure 5.

The autoencoder attempts to minimize the reconstruction error. The encoder defines an aggregated posterior distribution of $q(z)$ on the latent vectors as follows:

$$q(Z) = \int_X E(Z|X)p_d(X)dX \tag{12}$$

where, $E(Z|X)$ is encoding distribution, $p_d(X)$ is the data distribution. Meanwhile, the encoder ensures the aggregated posterior distribution $q(Z)$ can fool the discriminator into thinking that the latent comes from the true prior distribution $p(Z)$.

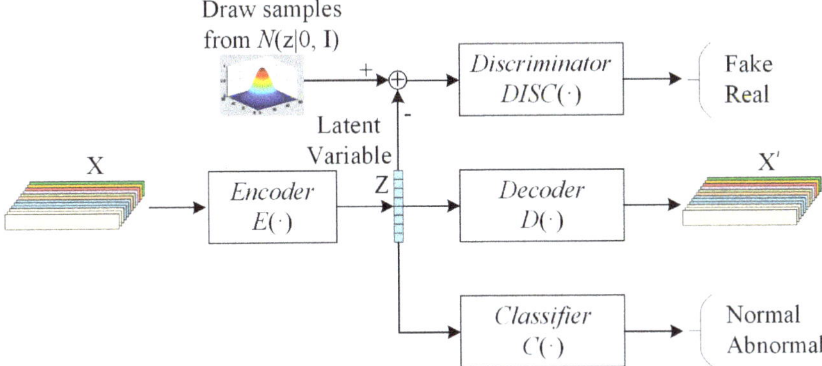

Figure 5. Framework of Semi-Supervised Deep Neural Network.

Figure 6 presents the detail architecture of the proposed network. Due to the samples are modeled as multi-channel 1D vectors, the encoder is equipped with 1D convolutional (Conv1D) layer. In detail, the encoder contains 4 Conv1D layers, each convolution layer contains 64 or 128 filters, and the kernel size is 5 and the stride is 2. The last layer of encoder is a fully connected layer without activation. The output dimension is related to latent space. In all experiments, 50 is the best choice for the dimension of latent space. The decoder contains three fully connected layers and a reshape layer to reconstruct samples. The output of the third fully connected layer is activated by sigmoid which related to the normalization of samples.

The discriminator also has three fully connected layers and the parameters are same as the decoder's fully connected layers. The difference is that the discriminator not only handles latent vectors, but also the samples drawn from $N(Z|0, I)$ which called Z_{real}. The discriminator is more like a function to measure the *similarity* between the latent vector and Z_{real}.

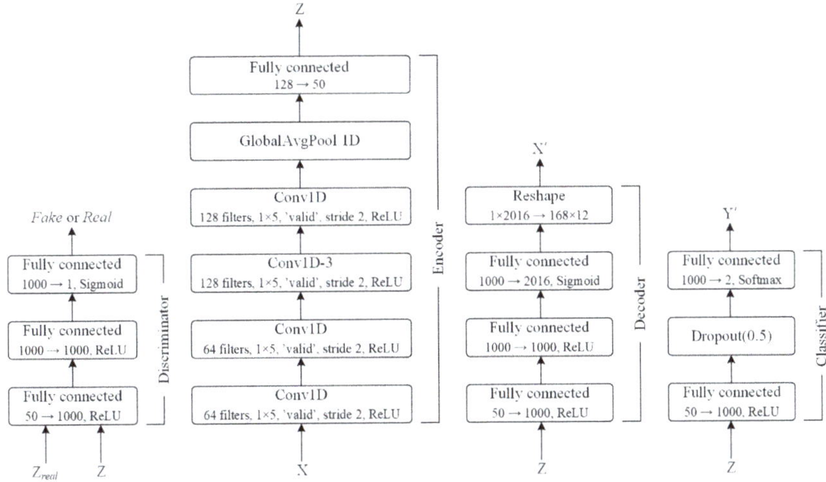

Figure 6. The detail architecture of the SSAE modules.

The classifier just contains two fully connected layers and a dropout layer. The last layer is activated by Softmax even if there are only two categories. The first fully connected layer is aimed to ascend dimension of latent features because there are difference in the customers' SM data. Ascending the dimension of latent features will improve the linear separability. The dropout layer is used to avoid overfitting. The second fully connected layer of classifier will find hyperplane between categories to complete classifying. In fact, the classifier in the SSAE is similar to SVM. However, we cannot use SVM to replace these 2 fully connected layers, because it is impossible to co-train SVM and autoencoder together. The separated training will lead to a decline in the learning efficiency, such as [23] could not get satisfactory performance of NTL detection.

4.2. Losses and Training

All modules of the SSAE are trained in thress phases:

1. **Reconstruction Phase**: The autoencoder updates the encoder and the decoder to minimize the reconstruction error:

$$min_{E,D}\mathbb{E}[||X - \tilde{X}||^2] \tag{13}$$

where $\tilde{X}$ is the reconstruction of X, and $|| \cdot ||^2$ is Euclidean distance.

2. **Regularization Phase**: Firstly, SSAE updates discriminator to apart the real samples from the encoded samples. In addition, then, SSAE updates encoder to confuse the discriminator. This phase can be represented by:

$$min_E max_{DISC}\mathbb{E}[log(DISC(Z_{real}))] + \mathbb{E}[log(1 - DISC(E(X)))] \tag{14}$$

3. **Classification Phase**: SSAE updates classifier and encoder Simultaneously by minimizing CrossEntropy and the distance of latent vectors within same class.

$$min_{E,C}\mathbb{E}[CrossEntropy(C(E(X)),y) + \omega(t) \cdot l_G] \tag{15}$$

where l_G is related to supervised feature clustering. It is defined as:

$$l_G((Z_i, y_i), (Z_j, y_j)) = \begin{cases} ||Z_i - Z_j||^2, & y_i = y_j \\ max(0, m - ||Z_i - Z_j||)^2, & y_i \neq y_j \end{cases} \tag{16}$$

where m is the margin between different classes. Due to the difference in the various customers' SM data, $l_G(\cdot)$ is designed to ascend distance of latent features between diferent catergaries with a minimum distance m. It also can be regarded as regularizer to classifier. Refer to [28], the weight ramp-up function $\omega(t)$ is defined as:

$$\omega(t) = exp[-5(1 - T)^2] \tag{17}$$

where, T increases linearly with the number of iterations from zero to one, in the first 40% (refer to [28]) of the total iterations.

The SSAE must be trained jointly with Adam [32]. The pseudocode of training algorithm with mini-batches is provided by Algorithm 1.

Algorithm 1 Mini-batch training of SSAE

Require: x = training inputs
Require: y = labels for labeled inputs in L
Require: z_{real} = random number from $N(0, I)$
Require: $E_\theta(x)$ = encoder with trainable parameters θ
Require: $D_\gamma(x)$ = decoder with trainable parameters γ
Require: $DISC_\phi(x)$ = discriminator with trainable parameters ϕ
Require: $C_\varphi(x)$ = classifier with trainable parameters φ
Require: $N(x)$ = stochastic input augmentation function
Require: $\omega(t)$ = weight of consistency loss
  1: **for** t = 1 to *iterations* **do**

  2:    Draw a minibatch B_u from unlabeled samples randomly
  3:    $\tilde{x}_i \leftarrow D_\gamma(E_\theta(x_i \in B_u))$
  4:    $z_i \leftarrow E_\theta(x_i \in B_u)$
  5:    $loss_{AE} \leftarrow \frac{1}{|B_u|} \sum_{i \in B_u} d(x_i, \tilde{x}_i)$

  6:    update θ *and* γ using Adam

  7:    $loss_{Disc} \leftarrow \frac{1}{|B_u|} \sum_{i \in B_u} \left[log(DISC_\phi(z_{real})) + log(1 - DISC_\phi(z_i)) \right]$

  8:    update ϕ *and* θ using Adam
  9:    Draw a balanced minibatch B_l from labeled samples randomly
10:    $\tilde{y}_i \leftarrow C_\varphi(E_\theta(x_i \in B_l))$
11:    Construct S, pairs of (x_i, x_j) with their labels, from B_l
12:    $loss_C \leftarrow \frac{1}{|B_l|} \sum_{i,j \in B_l} log\, \tilde{y}_i[y_i] + \omega(t) \cdot \frac{1}{|S|} \sum_{i,j \in S} l_G((E_\theta(x_i), y_i), (E_\theta(x_j), y_j))$
13:    update θ *and* γ using Adam
14: **end for**

---

5. Experiments and Discussion

5.1. Experiment Setting

5.1.1. Dataset

All training, validation and testing data are real data extracted from SGCC. This dataset contains 5000 three-phase four-wire industrial customers, where there are 461 normal and abnormal customers who have been inspected in-field manually. For each inspected customer, Hence, all normal and abnormal samples are labeled artificially based on inspection reports. The rule for data labeling is that all samples of same customer own the same label. The remaining unlabeled customers are randomly selected. All SM data are created as samples following the method mentioned in Section 3.4. Detailed information about the dataset is provided in Table 2. Although unlabeled customers could contribute more samples, we just select 500,000 samples randomly. The meters report electrical magnitudes listed in Table 1 every 15 min or 1 h. This paper unifies the frequency of all customers to 24 measurements/day.

Table 2. Brief information of dataset.

Type of NTL	Number of Customers	Number of Samples
Shunts of voltage	14	1321
Shunts of current	32	3609
Phase Shift	16	1798
Phase Disorder	25	2681
Phase Inversion	58	6374
Normal customer	316	35,506
unlabeled	4539	500,000

For each round of experiments, all samples are randomly splitted into training, validation and testing sets in approximated proportions of 10%, 10% and 80% by customers. It is worth noting that those three sub sets must follow above proportions to cover all types of NTL. Further, to verify generalization performance of algorithms, the samples of same customer would not be split.

5.1.2. Baseline

To demonstrate the effectiveness of our approach, we define several baseline methods for comparison. In the experiments, these methods are configured as:

1. **SVM**: The kernel is set as Radial Basis Function(RBF), penalty parameter is 0.01. Due to normal and abnormal are imbalance, we give them proper weight(normal:1, abnormal:2).
2. **KNN**: As same as [3], the best results were produced by KNN with 16 neighbors and euclidean distance.
3. **XGBoost**: The number of trees are 1200, the max depth of each tree is 7, minimum child weight equals 1 and the learning rate is configured as 0.01.
4. **MLP-3(MultiLayer Perceptron)**: Three fully connected layers, with 1000, 500 and 250 perceptrons and an additional classifier with 2 perceptrons. Between the 2nd layer and the 3rd, there is a dropout layer with a probability of 0.5. The first three layers all equipped l2 regulars and activated by ReLU. The learning rate is configured as 0.001.
5. **ResNet-20(Conv1D)**: It originates from [33] and we use 1D convolutional layers to replace all 2D convolutional layers. All parameters follow [33].

The SVM, KNN and XGBoost are based on scikit-learn [34]. The features feed to SVM, KNN and XGBoost are produced by Truncated Singular Value Decomposition(TruncatedSVD) from samples. The dimension of features configured as 50. MLP-3 and ResNet-20 are launched upon tensorflow. Further, they are trained on origin samples.

5.1.3. Metrics

Receiver Operating Characteristic (ROC) curve is a popular way to validate performance of classifier on imbalanced datasets and widely applied by [3,11,12,35]. It evaluates how fast the True Positive Rate (TPR) increases with the increase of the False Positive Rate (FPR). Commonly, AUC score, the area under the ROC curve, is used as primary metric. However, [36] mentioned that Precision-Recall Curve is the better choice rather than AUC score. [10] opted for the F1 score to evaluate the performance of algorithm. In order to compare with the above mentioned methods, we choose all of above metrics to completely evaluate the performance of the algorithm. Those metrics are defined as follows:

$$General\,Accuracy = \frac{TP + TN}{TP + FP + TN + FN} \tag{18}$$

$$TPR = \frac{TP}{TP + FN} \tag{19}$$

$$FPR = \frac{FP}{TN + FP} \tag{20}$$

$$Precision = \frac{TP}{TP + FP} \tag{21}$$

$$Recall = \frac{TP}{TP + FN} \tag{22}$$

$$F1 = \frac{2 \times Precision \times Recall}{Precision + Recall} \tag{23}$$

where TP(True Positives) is the number of NTL samples was correctly detected, FP(False Positives) are the number of NTL samples be classified as normal, and TN(True Negatives) means the number of normal samples be classified correctly, FN(False Negatives) represent the number of normal samples be classified as NTL samples. It should be noted that the decision threshold in *GeneralAccuracy*, *Precision*, *Recall*, and *F1* is 0.5.

5.2. Results

In this section, all experiments are repeated five times based on randomly split datasets (refer to Section 5.1.1), and the mean result will be provided.

5.2.1. Effect of Knowledge Embedded Sample Model

From Figure 4, we can get a rough impression that the linear separability of knowledge embedded sample is better than raw SM data. Table 3 presents further the detailed performance of raw SM data samples and knowledge-embedded samples. By comparing XGBoost and SSAE in both sample models, all metrics of both algorithms are improved obviously. It demonstrates that knowledge of electricity measurement is very helpful for NTL recognition. Particularly, knowledge embedded sample model plays a more important role to SSAE because it allows SSAE learning more advanced features from mass samples. On the contrary, SSAE could not learn any knowledge from raw SM data, and the performance of SSAE is similar to XGBoost.

Table 3. Comparison about sample embedded knowledge or not.

Methods	Precision	Recall	F1 Score	AUC Score	General Accuracy
XGBoost + raw SM data	0.911	0.538	0.676	0.917	0.905
XGBoost + knowledge embedded sample	**0.846**	**0.700**	**0.766**	**0.951**	**0.921**
ResNet-20 + raw SM data	0.913	0.578	0.708	0.916	0.909
ResNet-20 + knowledge embedded sample	**0.891**	**0.732**	**0.804**	**0.951**	**0.943**
SSAE + raw SM data	0.882	0.565	0.689	0.907	0.898
SSAE + knowledge embedded sample	**0.944**	**0.804**	**0.866**	**0.964**	**0.951**

5.2.2. Study of SSAE

We evaluate the performance of SSAE from the following aspects:

1. **Capability of semi-supervised learning**: As mentioned earlier, semi-supervised learning requires only a small number of labeled sample to complete the training of deep neural network and achieve ideal performance. Figure 7a provides the NTL detection effect of SSAE obtained with varying numbers of labeled samples. The results show that the SSAE can still obtain the F1 score of 0.775 and the AUC score of 0.938 with only 500 labeled samples. SSAE achieves the best results when the labeled sample reached 5000. Figure 8 further shows the feature learning ability of SSAE. By comparing two images, the overlapping regions of the features learned by SSAE in different categories become very small. It improves the linear separability of samples obviously.

2. **Effect of latent feature dimension**: The dimension of latent feature is a very important hyperparameter for SSAE. Figure 7b studied the effect of the dimension in the performance of SSAE. It can be seen from the results in Figure 7b, too small dimension will excessively cut down valid features, resulting in a decrease in the NTL detection performance. With the increase of the latent feature dimension, especially after more than 50, the effect will not continue to increase, but will decrease a little. The main reason is the latent features in larger dimension contains partial redundant information, which reduces its linear separability. By contrast, the best latent feature dimension is 50.

3. **Generalization performance**: In order to verify the generalization performance of SSAE, this paper launched the experiment with same parameters five times by randomly splitting the training set, validation set and testing set. The result of Figure 7c shows that the F1 score and the AUC score are between the ranges of [0.86, 0.883] and [0.965, 0.979], respectively. It is proved that

SSAE has achieved remarkable generalization performance. In order to reflect the fairness of the comparison, the median results of SSAE were selected in subsequent subsections for comparison.

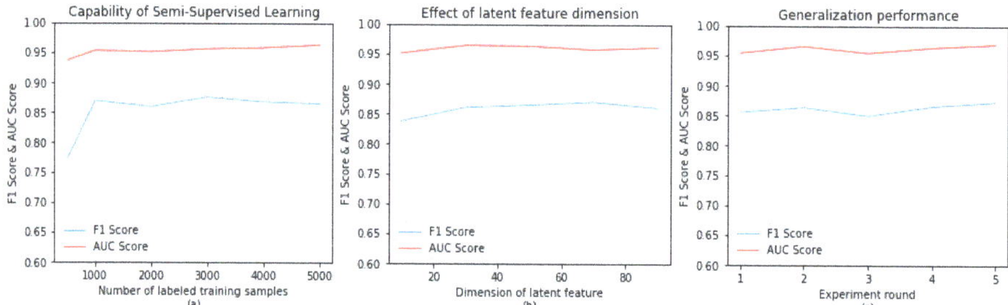

Figure 7. Study the performance of SSAE according to varying proportion of labeled samples, latent size and experiment round. (**a**) Capability of semi-supervised learning against NTL. (**b**) Studying the effect of latent size to SSAE. (**c**) F1 score and AUC score on each round of experiment.

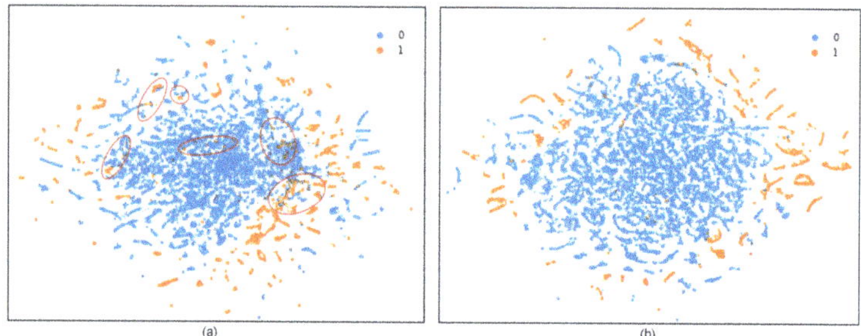

Figure 8. Comparison of t-SNE results of original samples and features learned by SSAE where 0 denotes normal samples plotted in blue, and 1 denotes NTL samples plotted in orange. (**a**) Original samples with embedding knowledge. (**b**) Latent features learned by SSAE.

4. **Convergence analysis**: For semi-supervised learning, the number of epochs is a very important parameter to avoid underfitting and overfitting. In this paper, the epoch is defined by training all labeled samples rather than unlabeled samples. Too small or too large an epoch value will lead to underfitting or overfitting, respectively. Figure 9 provides losses and scores with the epoch from 1 to 100. Between 40 and 60 epochs, losses and scores are relatively flat and stable. Before 40 epochs and after 60 epochs, there are large fluctuations. Especially after 60 epochs, the training loss continued to fall, the validation loss began to rise. At same time, the AUC score and the F1 score both decreased slightly, and SSAE is over-fitting obviously. Moreover, in Figure 9, the AUC score and F1 score reach 0.9738 and 0.8763 at the 50th epoch, respectively. Therefore, the epoch value of all experiments in this paper is fixed at 50.

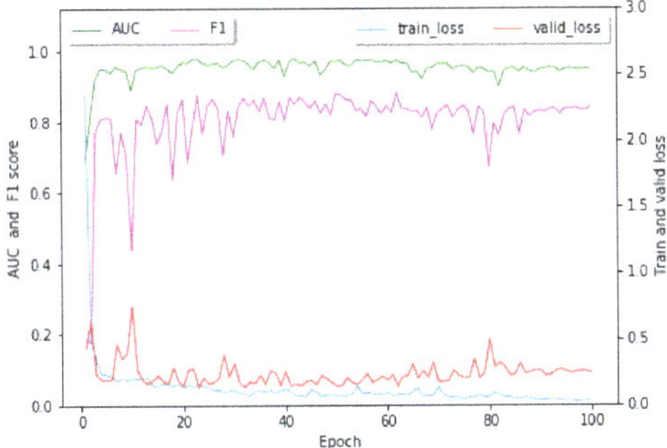

Figure 9. Convergence analysis of SSAE.

5.2.3. Comparison and Discussion

(1) Compared to Baselines

Table 4 and Figure 10 present the performance comparison of our approach and baselines. Overall, SSAE achieves the best results on all metrics. Benefit from the knowledge embedded sample model, SVM, KNN and XGBoost have also achieved good results, even using dimensionality reduction data as input features. Due to the stronger feature learning ability, MLP-3 and ResNet-20 is better than the above three methods. However, MLP-3 and ResNet-20 are supervised learning, and there can hardly avoid overfitting when labeled samples are extremely limited. In addition, the results of each of trial are very unstable. By using massive unlabeled samples and regularized losses, SSAE avoids overfitting successfully.

Table 4. NTL detection performance comparison (with knowledge).

Methods	Precision	Recall	F1 Score	AUC Score	General Accuracy
SVM	0.726	0.676	0.700	0.908	0.903
KNN	0.828	0.627	0.714	0.866	0.907
XGBoost	0.846	0.700	0.766	0.951	0.921
MLP-3	0.844	0.734	0.785	0.946	0.926
ResNet-20	0.891	0.732	0.804	0.951	0.934
SSAE	**0.944**	**0.804**	**0.866**	**0.964**	**0.951**

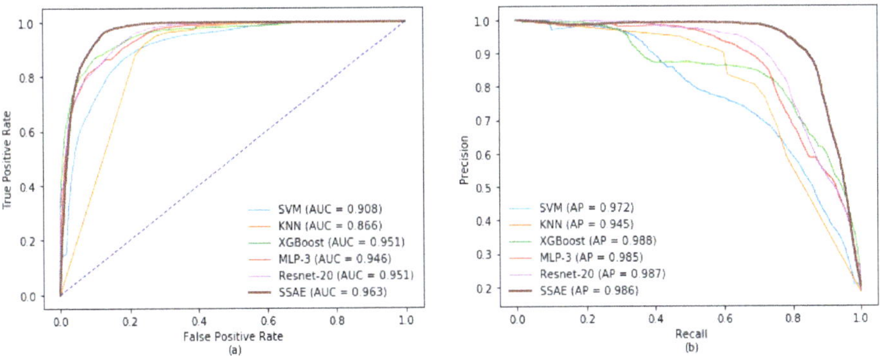

Figure 10. The ROC curve and PR curve of SSAE and baselines.

From Table 4, XGBoost obtained a good AUC score which close to the SSAE by selecting a very small decision threshold. Obviously, it will make the classifier be more sensitive, and lead to unstable. On the contrary, SSAE allows larger decision threshold without causing a significant drop in Precision and Recall. Refer to the results of SSAE presented in Table 4, Precision and Recall of SSAE outperform all baselines when the decision threshold is 0.5. It shows that SSAE separates normal and abnormal samples as much as possible. Hence, the classifier of SSAE will be more stable on varying scenarios.

(2) Comparison with other proposals

Table 5 shows a comparison between our method and the state-of-the-art approaches which report the AUC score and F1 score. It is worth mentioning that the AUC score or F1 score of state-of-the-art are referenced from their papers directly. Because our dataset cannot satisfy their requirements. For example, [35] needs observe meters as auxiliary data to help NTL detection. [3] requires GIS data, quality data and TECH data to achieve the best performance. [10,12] ask large number of labeled samples, and more than 1 year span of consumption data is [11]'s necessary condition.

Among these state-of-the-art approaches, [10,11,35] are based on artificial samples. The SMART attack model defined by [11] is the simplest situation because its fraud factor α_t is dominated by a fixed parameter. Due to this reason, [11] achieves the AUC score of 0.99. However, the realistic NTL is more similar to the adaptive attack model(FDI5) defined by [10]. As the key factor of the FDI5 is changed randomly and timely, [10] achieves the F1 score of 0.83 and [35] achieves the AUC score of 0.851. Even though their performance are poor enough refer to [11], their results are more convincing.

On the other hand, [3,12] and the SSAE are validated on realistic NTL samples. The results in the Table 5 show that the SSAE has achieved a large lead on AUC score and F1 score. The knowledge embedded sample model and deep semi-supervised learning are key reasons. Although [12] is also based on deep neural networks, its model is designed on electricity consumption completely and without any domain knowledge, so that its AUC score is not ideal. To avoid the limitation of information on electricity consumption, [3] supplements various auxiliary or privacy data to achieve notable improvement. It undoubtedly increases the difficulty of data acquisition, especially some data refer to customers' privacy. Our approach is a compromise solution which based on the SM data collected by the typical AMI system. It not only reduces the requirement of data types, but also protects customers' privacy. Besides knowledge embedded sample model, the SSAE has stronger feature learning and NTL detection capabilities. Even if raw SM data, SSAE still obtains an AUC score of 0.907.

Table 5. Comparison with the state-of-the-art.

Methods	AUC Score	F1 Score	NTL Sample	Dataset	Response Time
[10]	-	0.83	Artificial	Imbalance	3 weeks
[35]	0.851	-	Artificial	Imbalance	1 month
[11]	0.99	-	Artificial	Balance	1 year
[3]	0.91	-	Real	Imbalance	90 days
[12]	0.80	-	Real	Imbalance	4 weeks
SSAE	0.964	0.866	Real	Imbalance	1 week

6. Conclusions

This paper provides a novel knowledge embedded sample model and deep semi-supervised learning algorithm to detect NTL by using SM data. We first analyzed the characteristic of realistic NTL, and design a knowledge embedded sample model refer to the principle of electricity measurement. Next, we proposed an autoencoder based semi-supervised learning model. To avoid overfitting, we designed a regularization module, loss and training algorithm. Overall, our scheme outperforms all baselines and state-of-the-art results. In future work, it is promising to explore a new sample model and deep neural networks to adapt to possible public datasets.

Author Contributions: Conceptualization, X.L. and Y.Z.; methodology, X.L., Y.Z. and Y.Y.; validation, X.L. and L.F.; resources, L.F.; writing—original draft preparation, X.L., Z.W. and Y.Y.; writing—review and editing, F.W.; supervision, Z.W.

Funding: This research was funded by State Grid Jiangsu Electric Power Co., Ltd. Science and Technology Project: Research on Key Technologies of Big Customers Abnormal Electricity Consumption Diagnosis Based on Deep Adversarial Learning (grant number: J2019048).

Conflicts of Interest: The authors declare no conflict of interest.

References

1. Antmann P. Reducing technical and non-technical losses in the power sector. In *Background Paper for the WBG Energy Strategy*; Technical Report; The World Bank: Washington, DC, USA, 2009.

2. PR Newswire. World Loses $89.3 Billion to Electricity Theft Annually, $58.7 Billion in Emerging Markets. 2014. Available online: http://www.prnewswire.com/news-releases/world-loses-893-billion-to-electricity-theft-annually-587-billion-in-emerging-markets-300006515.html (accessed on 25 May 2019).

3. Buzau, M.-M.; Tejedor-Aguilera, J.; Cruz-Romero, P.; Gomez-Exposito, A. Detection of Non-Technical Losses Using Smart Meter Data and Supervised Learning. *IEEE Trans. Smart Grid* **2019**, *10*, 2661–2670. [CrossRef]

4. CEC. Monthly Statistics of China Power Industry. 2018. Available online: http://english.cec.org.cn/No.110.1737.htm (accessed on 25 May 2019).

5. Messinis, G.M.; Hatziargyriou, N.D. Review of non-technical loss detection methods. *Electr. Power Syst. Res.* **2018**, *158*, 250–266. [CrossRef]

6. Ahmad, T.; Chen, H.; Wang, J.; Guo, Y. Review of various modeling techniques for the detection of electricity theft in smart grid environment. *Renew. Sustain. Energy Rev.* **2018**, *82*, 2916–2933. [CrossRef]

7. Wu, S.; Ji, C.; Sun, G.Q. A Clustering Algorithm Based on CUDA Technology for Massive Electric Power Load Curves. *Electr. Power Eng. Teachnol.* **2018**, *37*, 5–70. (In Chinese)

8. Yang, X.; Zhang, X.; Lin, J.; Yu, W.; Zhao, P. A Gaussian-mixture model based detection scheme against data integrity attacks in the smart grid. In Proceedings of the 2016 25th International Conference on Computer Communication and Networks (ICCCN), Waikoloa, HI, USA, 1–4 August 2016; pp. 1–9.

9. Glauner, P.; Meira, J.A.; Dolberg, L.; State, R.; Bettinger, F.; Rangoni, Y. Neighborhood features help detecting non-technical losses in big data sets. In Proceedings of the 2016 IEEE/ACM 3rd International Conference on Big Data Computing Applications and Technologies (BDCAT), Shanghai, China, 6–9 December 2016; pp. 253–261.

10. Zanetti, M.; Jamhour, E.; Pellenz, M.; Penna, M.; Zambenedetti, V.; Chueiri, I. A Tunable Fraud Detection System for Advanced Metering Infrastructure Using Short-Lived Patterns. *IEEE Trans. Smart Grid* **2019**, *10*, 830–840. [CrossRef]

11. Messinis, G.M.; Rigas, A.E.; Hatziargyriou, N.D. A Hybrid Method for Non-Technical Loss Detection in Smart Distribution Grids. *IEEE Trans. Smart Grid* **2019**. [CrossRef]

12. Zheng, Z.; Yang, Y.; Niu, X.; Dai, H.-N.; Zhou, Y. Wide and deep convolutional neural networks for electricity-theft detection to secure smart grids. *IEEE Trans. Ind. Inform.* **2018**, *14*, 1606–1615. [CrossRef]

13. Jindal, A.; Dua, A.; Kaur, K.; Singh, M.; Kumar, N.; Mishra, S. Decision tree and SVM-based data analytics for theft detection in smart grid. *IEEE Trans. Ind. Inf.* **2016**, *12*, 1005–1016. [CrossRef]

14. Coma-Puig, B.; Carmona, J.; Gavalda, R.; Alcoverro, S.; Martin, V. Fraud detection in energy consumption: A supervised approach. In Proceedings of the 2016 IEEE International Conference on Data Science and Advanced Analytics (DSAA), Montreal, QC, Canada, 17–19 October 2016; pp. 120–129.

15. Nagi, J.; Yap, K.S.; Tiong, S.K.; Ahmed, S.K.; Mohamad, M. Nontechnical loss detection for metered customers in power utility using support vector machines. *IEEE Trans. Power Deliv.* **2009**, *25*, 1162–1171. [CrossRef]

16. Jokar, P.; Arianpoo, N.; Leung, V.C. Electricity theft detection in AMI using customers' consumption patterns. *IEEE Trans. Smart Grid* **2015**, *7*, 216–226. [CrossRef]

17. Ramos, C.C.O.; de Souza, A.N.; Falcao, A.X.; Papa, J.P. New insights on nontechnical losses characterization through evolutionary-based feature selection. *IEEE Trans. Power Deliv.* **2011**, *27*, 140–146. [CrossRef]

18. Monedero, I.; Biscarri, F.; León, C.; Guerrero, J.I.; Biscarri, J.; Millán, R. Detection of frauds and other non-technical losses in a power utility using Pearson coefficient, Bayesian networks and decision trees. *Int. J. Electr. Power Energy Syst.* **2012**, *34*, pp. 90–98. [CrossRef]

19. Krishna, V.B.; Weaver, G.A.; Sanders, W.H. PCA-based method for detecting integrity attacks on advanced metering infrastructure. In Proceedings of the 2015 12th International Conference on Quantitative Evaluation of Systems, Madrid, Spain, 1–3 September 2015; pp. 70–85.

20. Júnior, L.A.P.; Ramos, C.C.O.; Rodrigues, D.; Pereira, D.R.; de Souza, A.N.; da Costa, K.A.P.; Papa, J.P. Unsupervised non-technical losses identification through optimum-path forest. *Electr. Power Syst. Res.* **2016**, *140*, 413–423. [CrossRef]

21. Guerrero, J.I.; León, C.; Monedero, I.; Biscarri, F.; Biscarri, J. Improving knowledge-based systems with statistical techniques, text mining, and neural networks for non-technical loss detection. *Knowl. Based Syst.* **2014**, *71*, 376–388. [CrossRef]

22. Wei, L.; Keogh, E. Semi-supervised time series classification. In Proceedings of the 12th ACM SIGKDD International Conference on Knowledge Discovery and Data Mining, Philadelphia, PA, USA, 20–23 August 2006; p. 748.

23. Tacón, J.; Melgarejo, D.; Rodríguez, F.; Lecumberry, F.; Fernández, A. Semisupervised Approach to Non Technical Losses Detection. *Phys. Lett. B* **2014**, *378*, 698–705.

24. Russakovsky, O.; Deng, J.; Su, H.; Krause, J.; Satheesh, S.; Ma, S.; Huang, Z.; Karpathy, A.; Khosla, A.; Bernstein, M.; et al. ImageNet Large Scale Visual Recognition Challenge. *Int. J. Comput. Vis. (IJCV)* **2015**, *115*, 211–252. [CrossRef]

25. Deng, L.; Hinton, G.; Kingsbury, B. New types of deep neural network learning for speech recognition and related applications: An overview. In Proceedings of the 2013 IEEE International Conference on Acoustics, Speech and Signal Processing, Vancouver, BC, Canada, 26–31 May 2013; pp. 8599–8603.

26. Wu, Y.; Schuster, M.; Chen, Z.; Le, Q.V.; Norouzi, M.; Macherey, W.; Krikun, M.; Cao, Y.; Gao, Q.; Macherey, K.; et al. Googles Neural Machine Translation System: Bridging the Gap between Human and Machine Translation. *arXiv* **2016**, arXiv:1609.08144.

27. Makhzani, A.; Shlens, J.; Jaitly, N.; Goodfellow, I.; Frey, B. Adversarial autoencoders. *arXiv* **2015**, arXiv:1511.05644.

28. Laine, S.; Aila, T. Temporal Ensembling for Semi-Supervised Learning. *arXiv* **2017**, arXiv:1610.02242.

29. Luo, Y.; Zhu, J.; Li, M.; Ren, Y.; Zhang, B. Smooth Neighbors on Teacher Graphs for Semi-Supervised Learning. In Proceedings of the IEEE/CVF Conference on Computer Vision and Pattern Recognition, Salt Lake City, UT, USA, 18–22 June 2018; pp. 8896–8905.

30. Wang R.C. Influence of Distribution Network Three-phase Unbalanceon Line Loss Increase Rate and Voltage Offset. *Electr. Power Eng. Teachnol.* **2017**, *36*, 131–136. (In Chinese)

31. Maaten, LV.D.; Hinton, G. Visualizing data using t-SNE. *J. Mach. Learn. Res.* **2008**, *9*, 2579–2605.

32. Kingma, D.; Ba, J. Adam: A Method for Stochastic Optimization. *arXiv* **2014**, arXiv:1412.6980.

33. He, K.M.; Zhang, X.; Ren, S.; Sun, J. Deep residual learning for image recognition. In Proceedings of the IEEE Conference on Computer Vision and Pattern Recognition, Las Vegas, NV, USA, 26 June–1 July 2016; pp. 770–778.

34. Pedregosa, F.; Varoquaux, G.; Gramfort, A.; Michel, V.; Thirion, B.; Grisel, O.; Blondel, M.; Prettenhofer, P.; Weiss, R.; Dubourg, V.; et al. Scikit-learn: Machine learning in Python. *J. Mach. Learn. Res.* **2011**, *12*, 2825–2830.

35. Zheng, K.; Chen, Q.; Wang, Y.; Kang, C.; Xia, Q. A Novel Combined Data-Driven Approach for Electricity Theft Detection. *IEEE Trans. Ind. Inf.* **2018**, *15*, 1809–1819. [CrossRef]

36. Coma-Puig, B.; Carmona, J. Bridging the Gap between Energy Consumption and Distribution through Non-Technical Loss Detection. *Energies* **2019**, *12*, 1748. [CrossRef]

Article

An Assessment of the Condition of Distribution Network Equipment Based on Large Data Fuzzy Decision-Making

Ning Wang and Fei Zhao *

School of Electrical and Electronic Engineering, North China Electric Power University, Baoding 071000, China; wangning@ncepu.edu.cn
* Correspondence: zhaofei@ncepu.edu.cn; Tel.: +86-312-752-2909

Received: 29 November 2019; Accepted: 28 December 2019; Published: 1 January 2020

Abstract: As one of the most important components of power grid, a distribution network is the most vulnerable part in the face of various uncertainties, and influences the stability and economy of a power system. In this paper, the operational information, hardware information and human factors were considered, and a state evaluation model of multi-source information fusion was established. Based on big data fuzzy iteration method and a weighted expert library, a weighted distribution of multi-source information was obtained, and an equipment condition assessment was carried out reasonably. Taking the distribution transformer as an example, the assessment showed that fusion of multi-source information presented in this paper is more comprehensive, and has the ability to reflect the state of equipment. The method proposed in this paper can accurately judge the running state of distribution equipment based on all kinds of information, and provides a reference for the follow-up power marketing for the status assessment of the user equipment.

Keywords: distribution network equipment; condition assessment; multi information source; fuzzy iteration

1. Introduction

With the increasing demand of power and the gradually expanding scale of power grids, higher requirements have been placed on the power supply reliability of systems. In order to improve the reliability of power supply, it is one of the most important tasks to analyze the operational status of a distribution network, and evaluating the status of distribution equipment for the distribution network is an important part of the transmission and distribution of electric energy in a power system [1–3]. Testing the distribution transformers, circuit breakers and other power distribution equipment, and evaluating its operating status, can ensure the safety performance of the equipment and the reliability of the distribution network, and it is of great significance to ensure the safe and stable operation of the distribution network and improve the economics of the power supply enterprise.

Due to the wide distribution and large amount of distribution equipment, and the large amount of operational monitoring data without uniform evaluation standards, great difficulties have been brought to the assessment of distribution equipment [4–10]. As one of the most important pieces of equipment of a distribution network, power transformers have also been paid much attention.

In response to the above problems, domestic and foreign experts have conducted a lot of research, and fuzzy evaluation, artificial intelligence and other methods are widely used in transformer state evaluation [11–15]. Reference [3] presents an evidential reasoning (ER) approach to the transformer condition assessment [3]. The methodology of transferring the transformer condition assessment problem into a multiple-attribute decision-making (MADM) solution under an ER framework is then presented in [3]. Based on the outputs of the ER approach, system operators can obtain an overall evaluation of an observed unit's condition; also, several units may be ranked in order of severity for system maintenance purposes [3]. In [16] they used artificial neural networks to construct a multi-information fusion model to comprehensively evaluate transformer status. The validity of the method was verified by a case study. Based on the fuzzy theory, references [17–20] evaluated the operating state of the transformer, and verified the effectiveness of the evaluation method by example. However, no specific standard was given for the selection of the evaluation index. In reference [21], the transformer evaluation index function was established based on the semi-Cauxi distribution, and the transformer state evaluation model of multi-information fusion was established considering the difference of the initial values of the indicators. In reference [22], an adaptive evolutionary limit learning machine algorithm is proposed, which was applied to the transformer state evaluation process, but the selection of transformer evaluation information needs to be optimized. In reference [23], based on the matter-element differential transformation, a transformer hierarchical evaluation model is established. At the same time, the concept of expert validity material element is introduced to make the weight determination more reasonable. In reference [24], based on DGA and SVM, the transformer state is evaluated, and the oil-immersed transformer is taken as an example for verification. Reference [25] introduces a transformer fleet monitoring solution to help the end user to group transformer assets and react accordingly to monitored situations [25]. Reference [26] establishes an index assessing system, considering the main body, the bushing and the accessories components, employs a Cauchy membership function for fuzzy grades division and represents a fuzzy evidence fusion method to handle the fuzzy evidence fusion processes [26]. In reference [27], a new multi-criterion based fuzzy logic model has been proposed to determine the overall health index of transformers. The method relies on the concentrations of individual dissolved gasses, significant diagnostic test results of transformer oil and paper insulation [27]. In reference [28], a novel method for DGA and FRA results unification is proposed, which is based on fuzzy sets application in failures detection and interpretation stages. In reference [29], a fuzzy logic technique for on-line condition diagnostics of transformer oil on the basis of leakage current flows through silica gel of breather and changes the color of silica gel [29].

The above evaluation methods consider the basic parameters of the transformer, operational data and other factors, but the evaluation criteria are greatly influenced by subjective factors, such as the experience of evaluating individuals and experts. Therefore, it is important to determine an evaluation method that is more in line with the actual operating conditions of the power distribution equipment, which can provide assistance for the state assessment of the distribution transformer, and have reference value for the economics of the power supply enterprise and the reliability of the power marketing.

In this paper, based on the large amount of operational information generated during the operation of the distribution network equipment, a state evaluation model of the distribution network equipment integrating multi-source information is established. The model comprehensively considers the critical state quantities of the distribution network equipment, and based on the fuzzy iterative method of big data and the establishment of the weight expert database, weights the multi-source information and reasonably evaluates the equipment status. Finally, taking the distribution transformer as an example, the evaluation results of the fusion of multi-source information proposed in this paper are proved to be more comprehensive. The method proposed in this paper can accurately judge the running status of the power distribution equipment based on various types of information, and provide a reference for the subsequent power marketing evaluation of the user equipment state, which is more instructive.

2. Feature Extraction and Scoring of Key State Quantities of Distribution Equipment

During the operation of the distribution network equipment, a large number of data are generated, including real-time data, historical data, hardware information, environmental conditions, etc. Therefore, appropriate processing is required to extract key state quantities and establish reasonable evaluation criteria.

2.1. Selection of the State Quantity of a Distribution Transformer

The state quantity of a distribution network device can directly or indirectly characterize various conditions during the operation of the equipment, and has guiding significance for the state evaluation of the distribution network equipment. At present, there are uniform standards and clear specifications in technology [30], as shown in Table 1.

Table 1. Typical state quantity of a distribution transformer.

Component	State Quantity	Reflected State
Winding and bushing	DC resistance	DC resistance exceeds the range.
	Insulation resistance	Insulation resistance is not normal.
	Temperature	The temperature of the joint is abnormal and The temperature rise is abnormal.
	Load rate	Overload.
	Degree of contamination appearance integrity	Severely contaminated or rusted appearance. damaged appearance.
	The temperature of respirator	Exceed the factory defaults.
	Three-phase unbalance rate	Three-phase unbalance rate is not normal.
Tap changer	Performance	Operation is not Normal.
Cooling system	Mechanical properties	Dry change fan vibration is not normal.
	Temperature	Temperature control device is abnormal.
Tank	Ground distance of the bench	The distance to the ground is not enough.
	Sealing	Finishing seal aging.
	Oil level	Oil level is not normal.
	Oil temperature	Oil temperature is abnormal.
Non-electricity protection device	Insulation resistance	Unqualified insulation.
Ground wire	Exterior	Insufficient connection or insufficient depth of grounding body.
Insulation	Grounding resistance	Grounding resistance is abnormal.
	Withstand voltage test	Pressure resistance is unqualified.
Identification	Identification integrity	Equipment identification is vague, incomplete, wrong, etc.

It can be seen from Table 1 that the state of the distribution transformer is large, and it is of great significance to reasonably select the state quantity and establish a scientific and comprehensive evaluation system for the state evaluation of the transformer. Since the distribution transformers used by industry and large users are mainly step-down transformers, and most of them are oil-immersed transformers, refer to the standards such as the State Network Distribution Equipment Status Evaluation Guidelines [31], according to the selection of key state quantities. The principle selects and classifies the state quantities of the distribution transformers in Table 1, as shown in Table 2.

Table 2. Selection and classification of distribution transformer key state quantity.

Classification	Specific Parts	State Quantity
Hardware situation	Sealing means	Sealing ability μ_1
	Degree of insulation	Withstand voltage test μ_2
	System contamination	Contamination μ_3
	Non-electricity protection device	Insulation resistance μ_{10}
	Winding and bushing	DC resistance μ_{11}
Operational situation	Oil level	Oil level μ_4
	Winding and bushing outer temperature	Temperature μ_5
	Grounding condition	Grounding down conductor appearance μ_6
	Respirator	Respirator status μ_7
	Load situation	Load rate μ_{12}
	Three-phase load balancing	Three-phase unbalance rate μ_{13}
Human factors	Equipment identity	Completeness of identification μ_8
	Tap changer	Tap changer performance μ_9

2.2. Scoring Criteria for Key State Quantities of Distribution Transformers

After the state quantity is selected, it needs to be scored to further evaluate the state of the distribution transformer. According to the unified regulations, the evaluation principles of each state quantity are shown in Table 3.

Table 3. Grading standard for distribution transformer status.

Serial Number	State Quantity Name	Evaluation Standard Description of Status Quantity Evaluation
1	Withstand voltage test	Whether the withstand voltage test is qualified or not.
2	Winding DC resistance	(1) The difference between the three phases of A, B and C is not more than 2% of the average value; when no neutral point is taken out, the value is 1%; (2) The relationship between the resistance values of the three phases is consistent with the factory.
3	Insulating oil	The degree of pressure resistance.
4	Insulation resistance	Below 20 °C, no less than 300 MΩ; less than 30% change from the previous time.
5	Equipment identification plate appearance	Whether the appearance is normal or not.
6	Sealing performance	Whether there is oil leakage or oil dropping.
7	Oil level	Whether the oil is abnormal.
8	Respirator performance	Whether the respirator is normal.
9	Grounding condition	Ground resistance cannot be greater than a specific value.
10	Oil temperature	Temperature value.
11	Three-phase unbalance rate	Percentage.
12	Load condition	Refer to the rated capacity to determine whether it is overload.
13	Casing contamination	Score according to the degree of contamination.
14	Temperature control system	Whether the temperature control system is normal.
15	Tap changer	Tap changer.
16	Non-electricity protection device	Whether the insulation is Qualified.
17	Environmental temperature and humidity information	Refer to the transformer equipment manual and determine it according to the temperature and humidity standards of the reference manual.
18	Operation hours	Years from the time of commissioning.
19	Family quality defect	Score according to no defects, potential defects, influential defects, and fatal defects.
20	Similar equipment failure rate	Score according to the probability of failure rate
21	Equipment maintenance record	Whether the equipment has ever failed, whether it has been overhauled.

It can be seen from Table 3 that the state quantities of the transformer have both qualitative indicators and quantitative indicators of different orders of magnitude and dimensions, so the state quantities need to be normalized before evaluation. The state quantities, including winding DC resistance, oil temperature and other state quantities, which can make the state of the equipment better when become smaller or lower, are treated by Equation (1); the state quantities such as withstand voltage test, insulation resistance and other state quantities, which can make the state of the equipment better when it becomes more powerful and larger, is treated by Equation (2); for state quantities of qualitative measurements (running time, containment performance, etc.), the degree of deterioration is given empirically based on experience.

$$\mu_{ij} = \begin{cases} 0 & \mu_{ij} \leq \mu_{ij0} \\ \frac{\mu_{ij}-\mu_{ij0}}{\mu_{ij1}-\mu_{ij0}} & \mu_{ij0} < \mu_{ij} \leq \mu_{ij1} \\ 1 & \mu_{ij} > \mu_{ij1} \end{cases} \tag{1}$$

$$\mu_{ij} = \begin{cases} 1 & \mu_{ij} < \mu_{ij1} \\ \frac{\mu_{ij0}-\mu_{ij}}{\mu_{ij0}-\mu_{ij1}} & \mu_{ij1} \leq \mu_{ij} < \mu_{ij0}, \\ 0 & \mu_{ij} \geq \mu_{ij0} \end{cases} \tag{2}$$

where $\mu_{ij}(i = 1, 2, \cdots, 9)$ is the value of j is determined by the state quantity; i indicates the relative deterioration degree of the state quantity, and the value range is [0,1]; μ_{ij} indicates the observation value; / indicates the ideal value or the factory value; μ_{ij1} indicates the attention value or the warning value. The value of μ_{ij0}, μ_{ij1} refers to reference [32,33].

According to the state quantity evaluation criteria given in Table 3, with reference to [32,33], and combined with the experience of a large number of experts and long-term experience, the evaluation set of the key state quantities of the distribution transformer in Table 2 is shown in Table 4.

Table 4. Evaluation set of distribution transformer key state quantity.

State Quantity	Description	Evaluation Set				
		Excellent	Good	General	Malfunction	Serious Failure
Sealing performance μ_1	Oil leakage situation $\mu_{1,1}$	0.2	0.2	0.3	0.2	0.1
	Oil dripping situation $\mu_{1,2}$	0	0	0.1	0.1	0.8
	Oil spilling situation $\mu_{1,3}$	0	0	0	0	1
Withstand voltage test μ_2	Pressure resistance $\mu_{2,1}$	0	0	0	0.1	0.9
Contamination μ_3	A small amount of contamination $\mu_{3,1}$	0.9	0.1	0	0	0
	More pollution $\mu_{3,2}$	0.8	0.1	0.1	0	0
	Obviously damaged rust $\mu_{3,3}$	0.1	0.2	0.3	0.3	0.1
	Severely contaminated and blocked $\mu_{3,4}$	0	0	0.2	0.5	0.3
Oil level μ_4	Oil level gauge indicates abnormality $\mu_{4,1}$	0.1	0.2	0.3	0.2	0.2
	Oil level gauge no indication $\mu_{4,2}$	0.1	0.1	0.3	0.3	0.2
Temperature μ_5	Temperature of connector is too high $\mu_{5,1}$	0.1	0.3	0.4	0.2	0
	Rise of temperature is not normal $\mu_{5,2}$	0.1	0.2	0.3	0.3	0.1
Grounding down conductor appearance μ_6	Lack of connection $\mu_{6,1}$	0.1	0.2	0.3	0.3	0.1
	Insufficient depth $\mu_{6,2}$	0.2	0.3	0.4	0.1	0
Respirator condition μ_7	The respirator is completely discolored by moisture $\mu_{7,1}$	0.3	0.3	0.3	0.1	0
	The respirator is completely breathless $\mu_{7,2}$	0.3	0.3	0.3	0.1	0
Identification integrity μ_8	Lack of identification $\mu_{8,1}$	0	0.1	0.2	0.5	0.2
	Wrong identifies or no identifies $\mu_{8,2}$	0	0	0.1	0.4	0.5
Tap changer performance μ_9	Tap position power indicates abnormal. $\mu_{9,1}$	0	0.5	0.5	0	0

3. Weight Determination Based on Fuzzy Iteration and Expert Weighted Database

After the key state quantity of the power distribution switch is selected, it is necessary to perform reasonable weight allocation for each state quantity to perform comprehensive evaluation of the state of the distribution network equipment. In this paper, we use the eclectic fuzzy decision-making and multi-level fuzzy comprehensive evaluation model to analyze the previous data of the distribution transformer; continuously update the weight ratio of the evaluation set through the weight inverse operation; reduce the influence of subjective factors brought by the expert review opinions; and improve the data, the reliability of the analysis and ultimately the establishment of a weight expert database.

3.1. Compromising Fuzzy Decision Weight Solving Process

The flow chart of the compromise fuzzy decision [34] is shown in Figure 1. The basic principle is the virtual fuzzy positive ideal and the fuzzy negative ideal. Then, the Euclidean distance method is used to determine the distance between the candidate object and the fuzzy positive and negative ideals, and the membership degree belonging to the fuzzy positive ideal is calculated to determine the selection scheme. The greater the degree of membership, the better the solution and the priority.

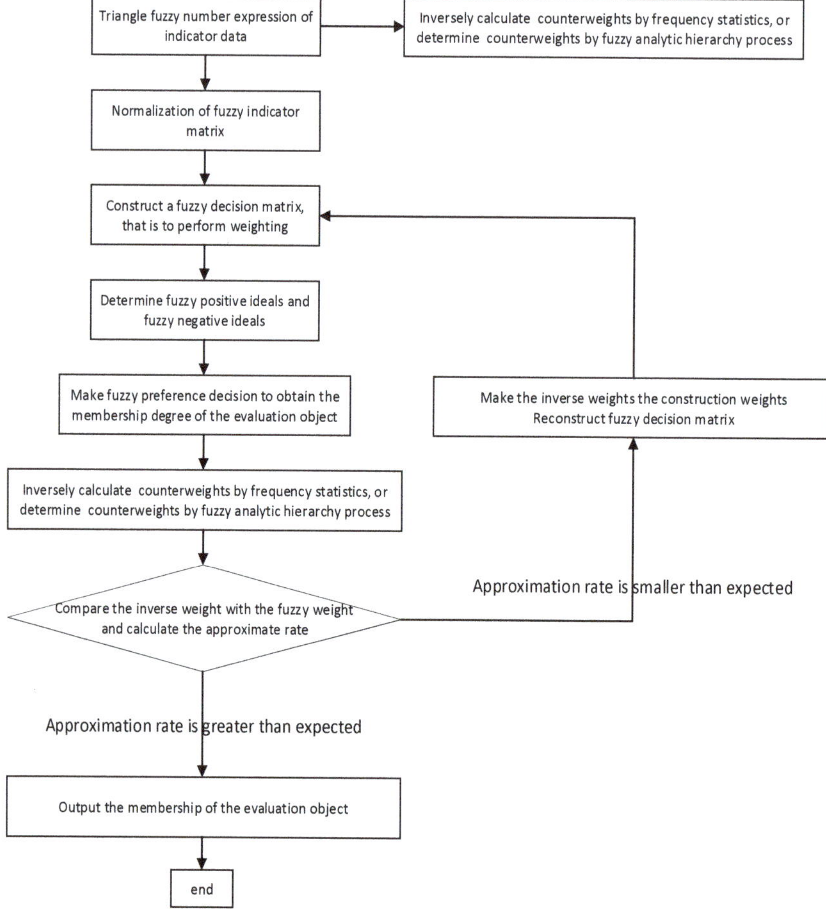

Figure 1. A flow chart of eclectic fuzzy decision-making model.

The basic solution steps for the compromising fuzzy decision are as follows:

Step 1: The indicator data is transformed into a triangle fuzzy number representation. Let $F(R)$ be the overall fuzzy set on R, set $M \in F(R)$. The membership function μ_M of M is expressed as

$$\mu_M(x) = \begin{cases} \frac{x-l}{m-l}, x \in [l, m] \\ \frac{x-u}{m-u}, x \in [m, u] \\ 0, x < l \text{ or } x > u \end{cases} ,$$ (3)

where $l \le m \le u$, and M is called a triangular fuzzy number, which is recorded as $M = (l, m, u) = (m_L, m, m_R)$.

According to the Equation (3), the qualitative index, the quantitative index and the weight data in the state quantity are unified into a triangular fuzzy number.

1. For the qualitative indicators $\mu_i(i = 1, 2, \cdots, 9)$ in the distribution transformer, they need to be converted into quantitative indicators according to Table 5.

2. The quantitative index value $\mu_i(i = 10, 11, \cdots, 13)$ for the critical state quantity of the distribution transformer needs to be written in the form of a triangular fuzzy number, as shown in Equation (4).

$$\mu_i = (\mu_i, \mu_i, \mu_i).$$ (4)

After all the indicators are converted into triangular fuzzy numbers, the fuzzy indicator matrix is obtained and recorded as $F = (f_{ij})_{m \times n}$.

3. The representation of the triangular fuzzy number of the weight vector. For the quantitative indicator, according to Equation (4), the triangular fuzzy number of its weight is expressed as follows:

$$w = [(w_1, w_1, w_1), (w_2, w_2, w_2), \cdots (w_i, w_i, w_i)].$$ (5)

For the weight of qualitative indicators, use the transformation method of Table 5 to convert it into an expression of triangular fuzzy numbers.

Table 5. Triangular fuzzy number ratio method for transforming qualitative index into quantitative index.

Quantitative Value Attributes	Cost Indicator	Profitability Indicator
(0,0,1)	Highest	Lowest
(1,1,2)	Very high	Very low
(2,3,4)	High	Low
(4,5,6)	General	General
(6,7,8)	Low	High
(7,8,9)	Very low	Very high
(9,10,10)	Lowest	Highest

Step 2: Normalize F. Suppose there are N evaluation objects, and the evaluation index $j(j \in N)$ corresponds to N fuzzy index values in F, and is denoted as $x_i = (a_i, b_i, c_i), (i = 1, 2, \cdots, N)$. Then, the normalization equation of x_i is as follows:

1. When x_i is the fuzzy indicator value corresponding to the cost indicator, the normalization equation is:

$$y_i = \left(\frac{\min(a_i)}{c_i}, \frac{\min(b_i)}{b_i}, \frac{\min(c_i)}{a_i} \wedge 1 \right).$$ (6)

2. When x_i is the fuzzy indicator value corresponding to the profitability indicator, the normalization equation is:

$$y_i = \left(\frac{a_i}{\max(c_i)}, \frac{b_i}{\max(b_i)}, \frac{c_i}{\max(a_i)} \wedge 1 \right). \tag{7}$$

The normalized fuzzy indicator matrix is recorded as $R = (y_{ij})_{m \times n}$.

Step 3: Construct a fuzzy decision matrix D. The fuzzy decision matrix can be obtained by weighting R:

$$D = (r_{ij})_{m \times n}, \tag{8}$$

where

$$r_{ij} = w \Theta y_{ij} (i = 1, 2, \cdots, N, j = 1, 2, \cdots, N) \tag{9}$$

Step 4: Determine the fuzzy positive ideal M^+ and the fuzzy negative ideal M^-.
Assume

$$M^+ = (M_1^+, M_2^+, \cdots, M_N^+)$$
$$M^- = (M_1^-, M_2^-, \cdots, M_N^-) \tag{10}$$

where $M_j^+ = \max\{r_{1j}, r_{2j}, \cdots r_{nj}\} (j = 1, 2, \cdots, N)$ and $M_j^- = \max\{r_{1j}, r_{2j}, \cdots r_{mj}\} (j = 1, 2, \cdots, N)$ respectively represent the fuzzy maximum and minimum values corresponding to the fuzzy index of column j in the fuzzy decision matrix.

Step 5: Determine the distance d_i^+, d_i^- between the evaluated object i and M^+, M^-.

$$d_i^+ = \sqrt{\sum_{j=1}^{N} (r_{ij} - M_j^+)^2}, i = 1, 2, \cdots, N. \tag{11}$$

$$d_i^- = \sqrt{\sum_{j=1}^{N} (r_{ij} - M_j^-)^2}, i = 1, 2, \cdots, N. \tag{12}$$

Step 6: Fuzzy optimal decision making. Let the evaluation object i be subordinate to the fuzzy positive ideal membership degree as μ_i, and then fuzzy optimal decision making. Let μ_i be the membership degree that the evaluation object i subordinate to fuzzy positive ideal; then

$$\mu_i = \frac{d_i^-}{d_i^+ + d_i^-}, i = 1, 2, \cdots, N. \tag{13}$$

Obviously $0 \le \mu_i \le 1$, if A_i is closer to M^+, the closer μ_i is to 1. The classification results of the membership degree are used to sort the pros and cons of the sample and form a fuzzy expert group commentary set of the multi-level fuzzy comprehensive evaluation model.

3.2. Multi-level Fuzzy Comprehensive Evaluation Model

In order to reduce the influence of subjective factors caused by expert experience and avoid errors caused by data redundancy or errors or omissions, this paper adopts a combination of eclectic fuzzy decision-making and multi-level fuzzy comprehensive evaluation to improve the evaluation accuracy of distribution transformer state assessment. The specific steps of the model are as follows:

Step 1: Determine the set of objects to be evaluated $X\{x_1, x_2, x_3, \cdots, x_k\}$; determining factor set $U = \{u_1, u_2, \cdots, u_n\}$; confirm the comment set $V = \{v_1, v_2, \cdots, v_n\}$.

Step 2: According to the factor set U and the comment set V, the evaluation matrix R_i is obtained.

$$R_i = \left\{ \begin{array}{cccc} r_{11}^{(i)} & r_{12}^{(i)} & \cdots & r_{1m}^{(i)} \\ \vdots & \vdots & & \vdots \\ r_{n_i 1}^{(i)} & r_{n_i 1}^{(i)} & \cdots & r_{n_i m}^{(i)} \end{array} \right\}. \tag{14}$$

Step 3: Make a comprehensive decision for each U_i. Let the weight of U_i be assigned as $A_i = \left(a_1^{(i)}, a_2^{(i)}, \cdots, a_{n_i}^{(i)} \right)$, and $\sum_{i=1}^{n_i} a_i^{(i)} = 1$. If R_i is a one-factor matrix, then the first-level evaluation vector is obtained as follows

$$A_i \times R_i = (b_{i1}, b_{i2}, \cdots, b_{in})\Delta B_i, \; i = 1, 2, \cdots, s. \tag{15}$$

Step 4: Think of each U_i as a factor, so U is a single factor set, and the single factor judgment matrix of U is:

$$R = \left(\begin{array}{c} B_1 \\ B_2 \\ \vdots \\ B_s \end{array} \right) = \left(\begin{array}{cccc} b_{11} & b_{12} & \cdots & b_{1m} \\ \vdots & \vdots & & \vdots \\ b_{s1} & b_{s2} & \cdots & b_{sm} \end{array} \right). \tag{16}$$

Each U_i is considered part of U, reflecting a certain attribute of U, which can be weighted according to their importance.

$$A = (a_1, a_2, \cdots, a_s). \tag{17}$$

The second-level fuzzy comprehensive evaluation model is obtained as follows:

$$B = A \times R = (b_1, b_2, \cdots, b_m). \tag{18}$$

If there are more factors in each sub-factor $U_i = (i = 1, 2, \cdots, s)$ you can continue to divide U_i.

Step 5: After obtaining the weight distribution, replace it with Step 3 in the basic solution step of the compromise fuzzy decision, that is, the establishment of the fuzzy weight, and then obtain the final computer expert library through repeated iterations. This not only gives a review of the computer expert library for the new data, but also expands the sample data for the computer expert library.

4. Case Analysis

This paper selects 16 distribution transformer data in a certain area to verify the proposed state assessment model.

4.1. Distribution Transformer Basic Parameters

According to Table 2, there are 13 key state quantities of distribution transformers, including four quantitative indicators and nine qualitative indicators. To reduce data redundancy and clearly show the accuracy of the algorithm, Table 6 shows the rating of the five major factor sets of the distribution transformer, including a quantitative indicator: winding DC resistance; and four qualitative indicators: sealing performance μ_1, insulation performance μ_2, grounding situation μ_6 and marking situation μ_8.

4.2. Status Assessment of Distribution Transformers

According to the basic principle of the compromised fuzzy decision, the qualitative indicators (excellent, good, qualified, unqualified) are firstly analyzed according to the quantitative indicators. The quantitative criteria are shown in Table 7.

Table 6. The original data of distribution transformer.

Distribution Number	Score of DC Resistance μ_{11}	Expert group's Fuzzy Score on State Quantity			
		Sealing Performance μ_1	Insulation Performance μ_2	Grounding Condition μ_6	Identification Situation μ_8
T_1	97	Excellent	Excellent	Good	Excellent
T_2	96	Excellent	Good	Excellent	Good
T_3	95	Good	Excellent	Qualified	Good
T_4	95	Excellent	Good	Failed	Good
T_5	94	Good	Excellent	Good	Failed
T_6	93	Excellent	Good	Failed	Good
T_7	93	Good	Excellent	Qualified	Qualified
T_8	93	Good	Excellent	Good	Qualified
T_9	92	Good	Good	Excellent	Qualified
T_{10}	92	Failed	Good	Qualified	Excellent
T_{11}	92	Failed	Qualified	Good	Excellent
T_{12}	91	Good	Good	Qualified	Excellent
T_{13}	90	Good	Qualified	Failed	Good
T_{14}	89	Qualified	Good	Good	Qualified
T_{15}	88	Good	Excellent	Failed	Good
T_{16}	86	Good	Qualified	Good	Qualified

Table 7. Quantitative standard for qualitative index of distribution transformers.

Grade	Excellent	Good	Qualified	Failed
Quantization fuzzy number	(85,90,100)	(75,80,85)	(60,70,75)	(50,55,60)

The initial triangle weight value is:

$$W = \begin{bmatrix} (0.5, 0.5, 0.5) \\ (0.2, 0.2, 0.2) \\ (0.15, 0.15, 0.15) \\ (0.1, 0.1, 0.1) \\ (0.05, 0.05, 0.05) \end{bmatrix}^{T} i.$$

The index information and the weight information are transformed into triangular fuzzy numbers, and the fuzzy index matrix F is obtained as follows.

$$F = \begin{pmatrix}
(96,96,96) & (85,90,100) & (85,90,100) & (75,80,85) & (75,80,85) \\
(95,95,95) & (85,90,100) & (75,80,85) & (85,90,100) & (65,70,75) \\
(95,95,95) & (70,80,85) & (85,90,100) & (50,60,65) & (65,70,75) \\
(94,94,94) & (85,90,100) & (75,80,85) & (75,80,85) & (75,80,85) \\
(93,93,93) & (75,80,85) & (85,90,100) & (75,80,85) & (65,70,75) \\
(93,93,93) & (75,80,85) & (50,60,65) & (85,90,100) & (75,80,85) \\
(92,92,92) & (85,90,100) & (75,80,85) & (65,70,75) & (75,80,85) \\
(92,92,92) & (75,80,85) & (85,90,100) & (85,90,100) & (65,70,75) \\
(92,92,92) & (75,80,85) & (75,80,85) & (85,90,100) & (75,80,85) \\
(92,92,92) & (50,60,65) & (75,80,85) & (85,90,100) & (65,70,75) \\
(91,91,91) & (50,60,65) & (65,70,75) & (75,80,85) & (85,90,100) \\
(90,90,90) & (85,90,100) & (75,80,85) & (65,70,75) & (85,90,100) \\
(89,89,89) & (75,80,85) & (65,70,75) & (50,60,65) & (85,90,100) \\
(89,89,89) & (50,60,65) & (75,80,85) & (85,90,100) & (75,80,85) \\
(88,88,88) & (85,90,100) & (75,80,85) & (65,70,75) & (75,80,85) \\
(87,87,87) & (75,80,85) & (85,90,100) & (75,80,85) & (65,70,75)
\end{pmatrix}.$$

The data in F is normalized to obtain the fuzzy decision matrix D. According to Equation (10), the fuzzy positive ideal M^+ and the fuzzy negative ideal M^- are obtained as:

$$
M^+ = \begin{bmatrix} (0.5, 0.5, 0.5) \\ (0.17, 0.20, 0.20) \\ (0.1275, 0.15, 0.15) \\ (0.085, 0.1, 0.1) \\ (0.0425, 0.05, 0.05) \end{bmatrix}^T
\quad
M^- = \begin{bmatrix} (0.4531, 0.4531, 0.4531) \\ (0.1, 0.1222, 0.1412) \\ (0.0750, 0.0917, 0.1059) \\ (0.05, 0.0611, 0.0706) \\ (0.03, 0.0389, 0.0441) \end{bmatrix}^T .
$$

The fuzzy double peak value and the membership percentage value obtained by the fuzzy optimization decision after multiple times of the cycle are shown in Table 8, and finally, the state evaluation value of the distribution transformer is obtained.

Table 8. Results of distribution transformer status assessment.

Distribution Number	Fuzzy Positive Ideal	Fuzzy Negative Ideal	Final Score
T_1	0.0167	0.1775	91.39
T_2	0.03	0.1683	84.88
T_3	0.0699	0.1495	82.86
T_4	0.0333	0.1608	76.27
T_5	0.0465	0.1495	75.08
T_6	0.0989	0.1253	72.88
T_7	0.0558	0.1495	72.8
T_8	0.0501	0.151	68.13
T_9	0.0525	0.141	68.03
T_{10}	0.1286	0.1036	63.01
T_{11}	0.1396	0.0737	60.87
T_{12}	0.0685	0.1457	55.88
T_{13}	0.1064	0.106	49.92
T_{14}	0.1377	0.0956	44.62
T_{15}	0.0838	0.1428	40.98
T_{16}	0.0896	0.1393	34.54

The fuzzy positive ideal and the fuzzy negative ideal are determined by analyzing a large number of transformers of the same type. This paper adopts a combination of eclectic fuzzy decision-making and multilevel fuzzy comprehensive evaluation, and the weights of quantitative indicators and qualitative indicators can be obtained by performing repeated fuzzy iterations. The weight ratio of the evaluation set is constantly updated through the weight inverse operation, reducing the influence of subjective factors brought by the expert review opinions and avoiding errors caused by data redundancy or errors or omissions, which improving the reliability of data analysis and promoting the establishment of the weight expert database, finally. The distances between the evaluated object and the fuzzy positive and negative ideals are determined, and the membership degree belonging to the fuzzy positive ideal is calculated. The greater the membership degree, the better the state of the transformer. The smaller the membership degree, the worse the state of the transformer, so it is necessary for the operation and maintenance personnel to pay attention to it and arrange the maintenance work in good time. The comparison and analysis of the relevant data in Tables 6 and 8 show that the final score of each transformer in Table 8 can truly reflect the actual operation status of the transformer in Table 6, which provides quantitative parameters for the evaluation of distribution equipment, and which is beneficial to the further focus and field evaluation of the equipment, providing help for maintenance and operation.

The analysis of practical examples shows that the evaluation method is concise and intuitionistic, and the evaluation conclusion not only reflects the state of a single transformer, but also facilitates the ranking of the overall state of the transformer, which provides reasonable suggestions for orderly arrangement of transformer maintenance.

5. Conclusions

Taking the distribution transformer as an example, this paper establishes a self-assessment model of the distribution equipment based on multi-source heterogeneous information fusion. The operation conditions and reliability assessment of the equipment can be obtained through the integration of data from a variety of sources, and maintenance is arranged according to the health state of the equipment, which can support the improvement of power supply reliability and the economy of power marketing.

This method reduces the influence of subjective factors brought by the expert review opinions; avoids errors caused by data redundancy or errors or omissions; and improves the reliability of data analysis and the accuracy of distribution equipment state assessment. The data of 16 distribution transformers in a certain area were selected, their status was evaluated. The effectiveness of the method was verified by a practical example.

Author Contributions: Conceptualization, N.W.; methodology, F.Z.; software, N.W.; validation F.Z.; formal analysis, F.Z.; investigation, F.Z.; resources, N.W.; data curation, N.W.; writing—original draft preparation, F.Z.; writing—review and editing, N.W.; supervision, N.W.; project administration, N.W.; and funding acquisition, F.Z. All authors have read and agreed to the published version of the manuscript.

Funding: This research was funded by the Fundamental Research Funds for the Central Universities (2015MS83).

Conflicts of Interest: The authors declare no conflict of interest.

References

1. Wang, S.-C.; Wang, C.-S. *Analysis of Modern Distribution System*, 1st ed.; Higher Education Press: Beijing, China, 2007.
2. Neto, A.C.; Da Silva, M.; Rodrigues, A.B. Impact of distributed generation on reliability evaluation of radial distribution systems under network constraints. In Proceedings of the 2006 International Conference on Probabilistic Methods Applied to Power Systems, Stockholm, Sweden, 11–15 June 2006; IEEE: Piscataway, NJ, USA, 2006; pp. 1–6.
3. Tang, W.H.; Spurgeon, K.; Wu, Q.H.; Richardson, Z.J. An evidential reasoning approach to transformer condition assessments. *IEEE Trans. Power Deliv.* **2004**, *19*, 1696–1703. [CrossRef]
4. Song, K.; Liu, R.-H.; Kang, Z.-J. Hybrid data mining based fault diagnosis for distribution systems. *J. Electr. Power Sci. Technol.* **2010**, *2*, 68–72.
5. Zhang, Y.Y.; Wei, H.; Liao, R.J.; Wang, Y.Y.; Yang, L.J.; Yan, C.Y. A New Support Vector Machine Model Based on Improved Imperialist Competitive Algorithm for Fault Diagnosis of Oil-immersed Transformers. *J. Electr. Eng. Technol.* **2017**, *12*, 830–839. [CrossRef]
6. Wang, T.; He, Y.G.; Li, B.; Shi, T.C. Fault Diagnosis of Transformer Based on Self-powered RFID Sensor Tag and Improved HHT. *J. Electr. Eng. Technol.* **2018**, *13*, 2134–2143.
7. Yuan, F.; Guo, J.; Xiao, Z.H.; Zeng, B.; Zhu, W.Q.; Huang, S.X. A Transformer Fault Diagnosis Model Based on Chemical Reaction Optimization and Twin Support Vector Machine. *Energies* **2019**, *12*, 960. [CrossRef]
8. Guo, C.-X.; Gao, Z.-X.; Zhang, J.-J.; Bi, J.-Q. IOT based transmission and transformation equipment monitoring and maintenance assets management. *J. Electr. Power Sci. Technol.* **2010**, *4*, 36–41.
9. Fang, J.K.; Zheng, H.B.; Liu, J.F.; Zhao, J.H.; Zhang, Y.Y.; Wang, K. A Transformer Fault Diagnosis Model Using an Optimal Hybrid Dissolved Gas Analysis Features Subset with Improved Social Group Optimization-Support Vector Machine Classifier. *Energies* **2018**, *11*, 1922.
10. Fotuhi-Firuzabad, M.; Rajabi-Ghahnavie, A. An analytical method to consider DG impacts on distribution system reliability. In Proceedings of the 2005 IEEE/PES Transmission & Distribution Conference & Exposition: Asia and Pacific, Dalian, China, 18 August 2005; IEEE: Piscataway, NJ, USA, 2005; pp. 1–6.
11. Zhang, Z.; Zhao, W.-Q.; Zhu, Y.-L.; Wu, Z.-L.; Yang, J. Power transformer condition evaluation based on support vector regression. *Electr. Power Autom. Equip.* **2010**, *4*, 81–84.
12. Stefan, T.; Sebastian, C.; Mohammad, D.; Andreas, M.; Mohammad, H.S.; Martin, S. Diagnostic Measurements for Power Transformers. *Energies* **2016**, *9*, 347.
13. Yuan, J.-S.; Zhang, L.-W.; Wang, Y.; Shang, H.-K. Study of transformers fault diagnosis based on extreme learning machine. *Electr. Meas. Instrum.* **2013**, *50*, 21–26.

14. Xie, H.-X.; Shi, L.-P.; Hui, Z.-Y. Research on immune clustering algorithm for transformers fault diagnosis. *Electr. Meas. Instrum.* **2012**, *49*, 15–18.

15. Sun, L.G.; Liu, Y.Y.; Zhang, B.Y.; Shang, Y.W.; Yuan, H.W.; Ma, Z. An Integrated Decision-Making Model for Transformer Condition Assessment Using Game Theory and Modified Evidence Combination Extended by D Numbers. *Energies* **2016**, *9*, 697. [CrossRef]

16. Ruan, L.; Xie, Q.-J.; Gao, S.-Y. Application of artificial neural network and information fusion technology in power transformer condition assessment. *High Volt. Eng.* **2014**, *40*, 822–828.

17. Wang, F.-Z.; Shao, S.-M. Fuzzy strategy on running state evaluation of oil-immersed power transformer. *Comput. Simul.* **2015**, *32*, 141–145.

18. Kari, T.-S.-J.; Gao, W.-S.; Liu, Y.-Q.; Chen, Y.-Q.; Hua, L. In Condition assessment of power transformer using fuzzy and evidential theory. In Proceedings of the International Conference on Condition Monitoring & Diagnosis, Xi'an, China, 25–28 September 2016.

19. Liao, R.-J.; Wang, Q.; Luo, S.-J. Condition assessment model for power transformer in service based on fuzzy synthetic evaluation. *Autom. Electr. Power Syst.* **2008**, *32*, 70–74.

20. Geetha, M.; Jovitha, J. Intuitionistic Fuzzy Expert System Based Fault Diagnosis Using Dissolved Gas Analysis for Power Transformer. *J. Electr. Eng. Technol.* **2014**, *9*, 2058–2064.

21. Liao, R.-J.; Huang, F.-L.; Yang, L.-J. Condition assessment of power transformer using information fusion. *High Volt. Eng.* **2010**, *36*, 1455–1460.

22. Lei, F. Study on Condition Assessment and Fault Diagnosis for Power Transformers Based on Big Data Analytics. Master's Thesis, Southwest Jiaotong University, Chengdu, China, 2016.

23. Yang, L.-X.; Yu, F.-W.; Bao, Y. Classification evaluation of transformer insulation condition based on matter-element theory. *Electr. Power Autom. Equip.* **2010**, *30*, 55–59.

24. Singh, J.; Kumari, P.; Kaur, K.; Swami, A.K. Condition assessment of power transformer using SVM based on DGA. In Proceedings of the 2016 Al-Sadeq International Conference on Multidisciplinary in IT and Communication Science and Applications (AIC-MITCSA), Baghdad, Iraq, 9–10 May 2016; pp. 1–5.

25. Eke, S.; Clerc, G.; Aka-Ngnui, T.; Fofana, I. Transformer Condition Assessment Using Fuzzy C-means Clustering Techniques. *IEEE Electr. Insul. Mag.* **2019**, *35*, 47–55. [CrossRef]

26. Tian, F.L.; Jing, Z.Z.; Zhao, H.; Zhang, E.Z.; Liu, J.F. A Synthetic Condition Assessment Model for Power Transformers Using the Fuzzy Evidence Fusion Method. *Energies* **2019**, *12*, 857. [CrossRef]

27. Ranga, C.; Chandel, A.K.; Chandel, R. Condition assessment of power transformers based on multi-attributes using fuzzy logic. *IET Sci. Meas. Technol.* **2017**, *11*, 983–990. [CrossRef]

28. Gonzales, J.C.; Mombello, E.E. Power Transformer Condition Assessment Using DGA and FRA. *IEEE Lat. Am. Trans.* **2016**, *14*, 4527–4533. [CrossRef]

29. Jaiswal, G.C.; Ballal, M.S.; Renge, M.M. Condition Monitoring of Breather for Transformer Health Assessment. In Proceedings of the 8th IEEE International Conference on Power Electronics, Drives and Energy Systems (PEDES), Chennaik, India, 18–21 December 2018.

30. State Grid Corporation of China. *Q/GDW645-2011 Guidelines for State Evaluation of Distribution Network Equipment*; China Electric Power Press: Beijing, China, 2011.

31. Miao, F.; Ren, J.-W.; Tang, G.-Q.; Wei, J.-J. Assessment Based on Gray Clustering and Evidence Synthesis. *High Volt. Appar.* **2016**, *52*, 50–55.

32. *DL/T 596—1996 Prevenive Test Code for Electric Power Equipment*; China Electric Power Press: Beijing, China, 1996.

33. *Q/GDW 168—2008 Regulations of Condition-Based Maintenance Test for Electric Equipment*; China Electric Power Press: Beijing, China, 2008.

34. Zhang, Z.-C.; Rao, C.-J.; Wang, C. A Sort of Eclectic Fuzzy Decision-making Model and Its Application. *Oper. Res. Manag. Sci.* **2005**, *5*, 33–35.

Article

Current Balancing Algorithm for Three-Phase Multilevel Current Source Inverters

Faleh Alskran * and Marcelo Godoy Simões *

Electrical Engineering Department, Colorado School of Mines, Golden, CO 80401, USA
* Correspondence: falskran@mines.edu (F.A.); msimoes@mines.edu (M.G.S.)

Received: 9 January 2020; Accepted: 12 February 2020; Published: 16 February 2020

Abstract: In high power, medium voltage applications, Current Source Inverters CSIs are connected in parallel to accommodate high DC currents. Using a proper multilevel modulation technique, parallel-connected CSIs can operate as a Multilevel CSI (MCSI). The most common modulation technique for MCSIs is the Phase-Shifted Carrier SPWM (PSC-SPWM). The proper operation of the MCSI requires each CSI modules to have the same average current flowing through its sharing inductors. In practice, the average currents of the CSI modules deviate from their nominal values. Therefore, current balancing mechanisms must be implemented. In the literature, several solutions have been proposed to tackle the current imbalance problem. Most of these solutions are based on altering the phase-shift or magnitude of the carrier waveforms of the PSC-SPWM. They require dedicated PI controllers and they are applicable to MCSIs with specific numbers of levels. This paper proposes a Current Balancing Algorithm (CBA) that can be implemented in any MCSI with any number of levels. The proposed CBA does not require any PI controllers, nor does it require any alteration to the PWM carrier waveforms. The CBA is implemented using a modified Level-Shifted SPWM (LS-PWM). The modified LS-SPWM is shown to produce lower THD and lower di/dt when compared to the PSC-SPWM. The CBA and modified LS-SPWM where implemented in a proof-of-concept lab prototype. The experimental results are presented for the five-level and seven-level cases.

Keywords: current balancing algorithm; level-shifted SPWM; medium-voltage applications; multilevel current source inverter; motor drives; phase-shifted carrier SPWM; STATCOM

---

1. Introduction

Current Source Inverters (CSIs) have been used as alternatives to Voltage Source Inverters (VSIs) in applications such as: motor drives [1], STATCOMs [2], High Voltage Direct Current (HVDC) transmission stations [3], and renewable energy conversion systems [4]. Compared to VSIs, CSIs have several advantages and drawbacks. Their advantages include inherent fault tolerance and voltage boosting capabilities. Their main drawbacks are larger ohmic losses in their DC inductors and heavy inductor cores. Nevertheless, ongoing research in the field of High Temperature Superconductors (HTS) has yielded promising results [5–7]. In [6], utilizing HTS coils was shown to reduce a generator's losses and weight by half and one third, respectively. In the future, access to HTS inductors, with virtually zero ohmic and core losses will make CSIs more attractive to a broader range of applications.

In high power, medium voltage applications, several VSIs or CSIs are connected in parallel to divide the high DC current evenly between them. For these applications, CSIs are more attractive because parallel-connected CSIs can operate as a Multilevel Current Source Inverter (MCSI). MCSIs generate higher quality current waveforms that require smaller output capacitors and lower switching frequency.

Over the past few decades, several attempts have been made to conceive an effective MCSI topology. Relying on the duality principle [8], most topologies require the DC input to be a constant current source. In most cases, either a DC voltage source connected to a large smoothing inductor [9,10]

or a current-controlled buck converter [11–14] is used as a constant current source. The most common three-phase MCSI is shown in Figure 1. It was first introduced by Xiong et al. [9] and has been the subject of several papers over the past few years [10–14]. In addition to a current-controlled buck converter or a large smoothing inductor at the input, this topology consists of M modules, for a $2M + 1$ level converter. Each module contains six controllable unidirectional switches, an upper inductor, and a lower inductor. The upper and lower inductors are referred to as sharing inductors and they are only rated for $1/M_{th}$ of the DC current, where the current-controlled buck converter or large smoothing inductor at the input must be rated for the full input current. A unidirectional controllable switch may be realized using a Gate Turn-Off Thyristor (GTO), an Integrated Gate-Controlled Thyristor (IGCT), a Reverse-Blocking Insulated-Gate Bipolar Transistor IGBT (RB-IGBT), or a regular IGBT connected, in series, to a diode, as shown in Figure 1.

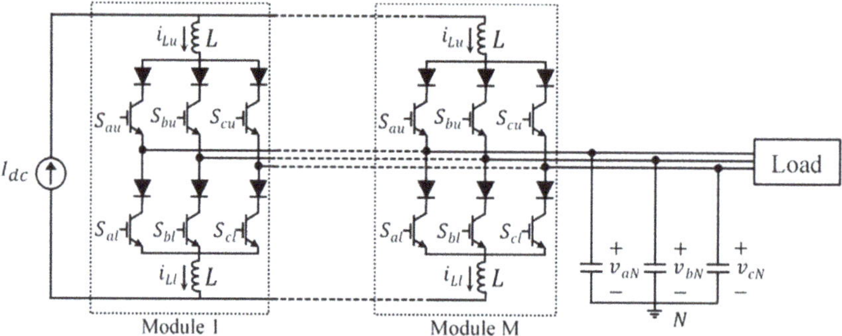

Figure 1. The multilevel current source inverter (MCSI) topology [9–14].

For proper operation of the MCSI, the values of the sharing inductors in each module must be identical. The main challenge in this topology is ensuring the main DC current is divided between the modules evenly. In [9], since a six-level staircase modulation was used, an optimized fixed switching sequence was proposed. The sequence minimizes the switching losses while ensuring the volt-second area remains close to zero for each inductor under steady-state conditions. Unfortunately, this method becomes very complex and unreliable when Pulse Width Modulation (PWM) techniques are used for a higher number of levels. Moreover, the number of levels achieved was six for a three-module converter; the zero level was not achieved using the staircase modulation.

In [11], a prototype of the same topology was built and tested. The input of the inverter, in this case, was a current-controlled buck converter. The modulation technique used was the Phase-Shifted Carrier Sinusoidal PWM (PSC-SPWM) [13]. The PSC-SPWM allows individual modules to be modulated independently using the so-called "tri-logic" SPWM [15]. To synthesize multilevel waveforms using the PSC-PWM, the phase-shift between any two adjacent module's carrier signals is set to $2\pi/M$. Although no current balancing technique was used, the average currents of the sharing inductors remained within a reasonable range of each other. Nonetheless, implementing such an inverter in a non-controlled environment, such as industrial applications, with no means to ensure a current-balanced operation is not safe. Disturbances and small differences in the electrical parameters between the modules can easily lead different inductors to have different average currents [11], and modules sustaining currents higher than their rated current could be damaged. The current imbalance problem is highlighted in [12]. Using the same modulation technique as that used in [11], the authors suggested two Proportional-Integral (PI) controllers-based solutions to ensure a current-balanced operation for the seven-level case. The first solution is to allow each module to use a small variation in the magnitude of carrier waveform as a control variable. The second solution is to allow the use of the phase-shift angle of the carrier waveform as a control variable. In both cases, to understand the input–output relationship, systems identification procedures were carried out. Both solutions were verified via simulation only, and both yielded satisfactory results. Another attempt to realize a current-balanced operation was presented in [14] where an improved version of the first solution proposed in [12] was implemented. The solution targets the seven-level case. It involved two dedicated PI controllers varying the magnitudes of the PSC-SPWM carrier waveforms. The effectiveness of the solution was verified via simulation and experimental results. However, there are several disadvantages to the proposed solutions [12,14]. By requiring multiple, dedicated, current-balancing, closed-loop control systems, the converter cost and complexity increases. Besides, the proposed alteration to the carrier waveforms has a negative impact on the Total Harmonic Distortion (THD).

This paper focuses on the voltage fed version of the MCSI. In the voltage fed MCSI, the modules are connected directly to the DC bus. Unlike the MCSI discussed in [9,10], a main smoothing inductor is not required here. This makes the MCSI modular and it reduces the number of required inductors. Also, unlike the MCSI considered in [11–14], the voltage fed MCSI does not need to be connected to the output of a current-controlled buck converter. Adding a buck converter to the input increases the cost, size, and complexity while reducing the boosting ratio of the MCSI. To compute the boosting ratio of the MCSI, the AC side power in Equation (1) is equated to the DC side power in Equation (2), assuming a lossless system. The resulting boosting ratio is given in Equation (3). In Equation (1), m is the modulation index of the MCSI, $\hat{V}_{LN}$ is the peak value of the line-to-neutral voltage on the AC side of the MCSI, and $\cos(\varphi)$ is the power factor. In Equation (2), d is the duty cycle of the current-controlled buck converter. The impact of d on the boosting ratio is evident in Equation (3). If the buck converter is eliminated, the boosting ratio can be derived by assuming $d = 1$. Therefore, eliminating the buck converter maximizes the boosting ratio.

$$P_{AC} = \frac{3}{2}\left(\frac{\sqrt{3}}{2} m I_{dc} \hat{V}_{LN} \cos(\varphi) \right) \tag{1}$$

$$P_{dc} = I_{dc} d V_{dc} \tag{2}$$

$$\frac{\hat{V}_{LN}}{V_{dc}} = \frac{4d}{3\sqrt{3}m\cos(\varphi)} \tag{3}$$

The main contribution of this paper is the introduction of a new Current Balancing Algorithm (CBA) to address the current imbalance issue in MCSIs. The CBA can be implemented in any MCSI regardless of the number of levels and it does not require any modification to the carrier's waveform, unlike the balancing methods in [12,14]. Moreover, the PWM technique used in this paper is a modified version of the Level-Shifted SPWM (LS-SPWM). It was first introduced to eliminate common-mode voltage in three-level Neutral-Point-Clamped (NPC) VSIs [16]. However, its implementation in MCSI has never been discussed or demonstrated in the literature. Compared to the widely used PSC-SPWM [13], the modified LS-SPWM produces lower THD and di/dt.

The rest of the paper is divided into four main sections. Section 2 presents the modified LS-SPWM. Section 3 explains how to solve the current-imbalance problem using the CBA. Section 4 presents experimental results verifying the CBA and modified LS-SPWM using a proof-of-concept prototype. The prototype has three modules. Therefore, it can operate as either a five-level MCSI, by only utilizing two modules, or a seven-level MCSI, by utilizing all three modules. Waveforms demonstrating the operation of the five-level and seven-level cases are presented. Finally, concluding remarks are presented in Section 5.

2. Modified LS-SPWM Suitable for MCSIs

An individual CSI must have one upper switch and one lower switch switched ON at any given instance [15]; this ensures that i_{Lu} and i_{Ll} will always have a circulation path. Therefore, the CSI has nine possible switching states. These states can be divided into three zero states and six active states, as illustrated in Figure 2. Zero states produce zero in all phases at the AC side while active states produce a positive current in one phase and a negative current in a different phase. It is important to note that any set of three-phase currents synthesized by the CSI using any switching state shown in Figure 2 will always add up to zero. Hence, any multilevel PWM technique must produce a set of three-phase currents that always adds up to zero. It has been was shown [16] that the Level-Shifted SPWM (LS-SPWM) does not satisfy this requirement. A simple solution is described in [16] for the three-level case. The same solution can be extended to any number of levels as follows:

(1)   Generate a set of three-phase modulated signals using the reference signals in Equations (4)–(6), m is the modulation index and M is the number of modules. The modulated signals should be generated by comparing Equations (4)–(6) to M level-shifted triangle waveforms.

(2)   A balanced set of three-phase currents, which add up to zero, can be obtained by computing Equations (7)–(9), where i_{1m}, i_{2m}, and i_{3m} are the modulated signals of Equations (4)–(6).

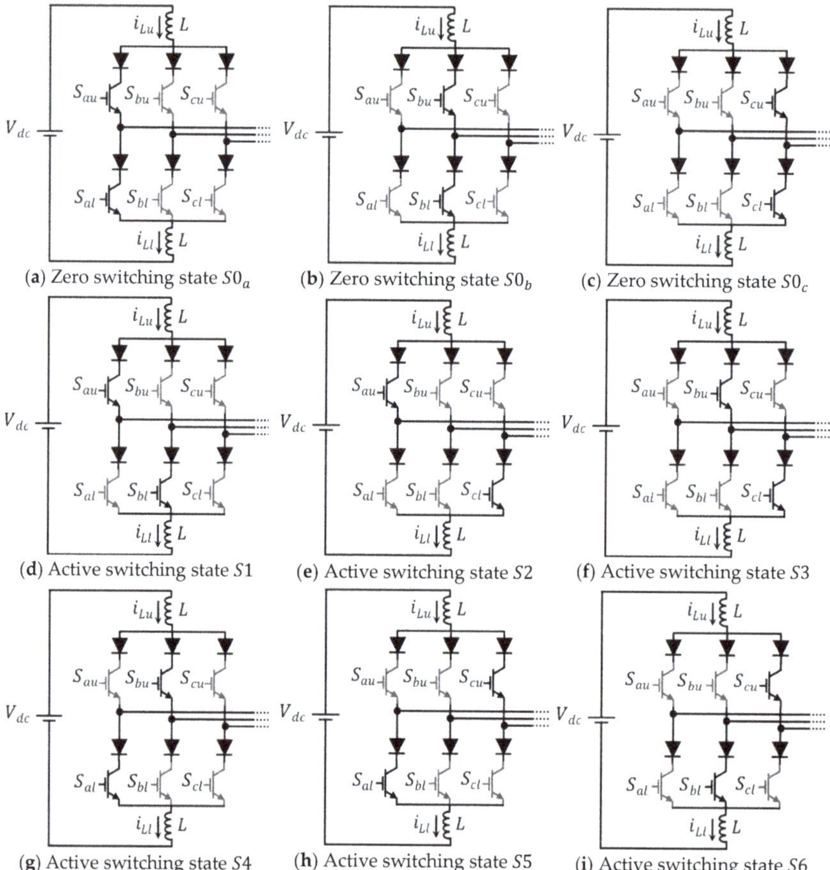

Figure 2. MCSI switching states: (**a–c**) Zero switching states, (**d–i**) Active switching states.

The obtained currents in Equations (7)–(9) have $2M + 1$ levels. Their magnitude is $\left(M\sqrt{3}\,m\right)/2$. To extend the upper limit of m to $2/\sqrt{3}$, instead of 1, a third harmonic component with a magnitude of $(-m)/6$ can be added to Equations (4)–(6) [17].

$$i_1 = \frac{M}{2}\left[m\cos\left(\omega t - \frac{\pi}{6}\right)\right] \tag{4}$$

$$i_2 = \frac{M}{2}\left[m\cos\left(\omega t - \frac{2\pi}{3} - \frac{\pi}{6}\right)\right] \tag{5}$$

$$i_3 = \frac{M}{2}\left[m\cos\left(\omega t + \frac{2\pi}{3} - \frac{\pi}{6}\right)\right] \tag{6}$$

$$i_{am} = i_{1m} - i_{2m} \tag{7}$$

$$i_{bm} = i_{2m} - i_{3m} \tag{8}$$

$$i_{cm} = i_{3m} - i_{1m} \tag{9}$$

The process described above is shown in Figure 3 for a seven-level case, $M = 3$. The modulated signals of Equations (4)–(6) are shown in Figure 3a while Equations (7)–(9) are shown in Figure 3b. For comparison, a set of three-phase currents were produced using the PSC-SPWM [10–14], as shown in Figure 4. The modified LS-SPWM resulted in a lower THD, 24.12% compared to 33.62% for the PSC-SPWM. Also, unlike the PSC-SPWM, the modified LS-SPWM results in modulated waveforms where the transition from one level to the next happens consecutively, i.e., the modulated waveforms increase or decrease by one level at a time. This results in a lower di/dt, which produces less Electromagnetic Interference (EMI).

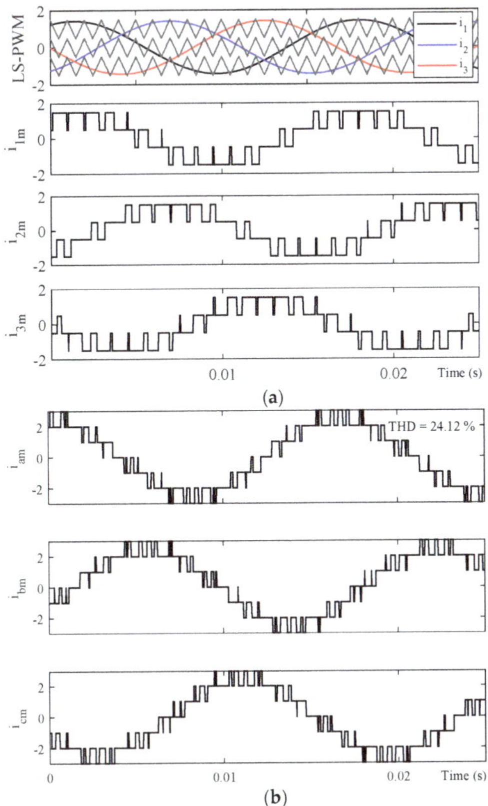

Figure 3. Modified LS-SPWM: (**a**)Three-phase currents obtained using the conventional LS-SPWM, $m = 0.95$ and $f_s = 1$ kHz, (**b**) Modified LS-SPWM, calculated using Equations (7)–(9).

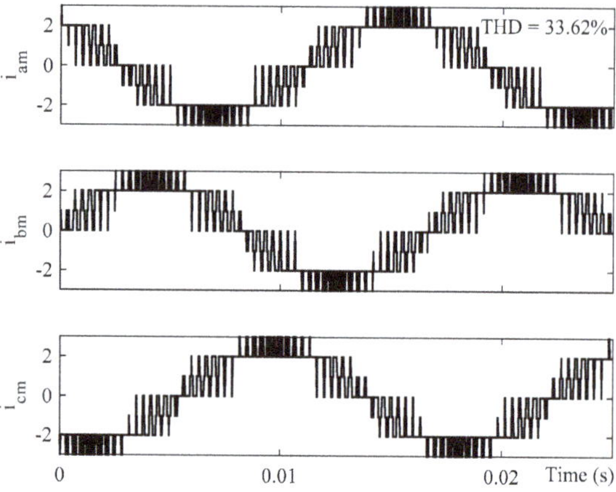

Figure 4. PSC-SPWM [13], $m = 0.95$ and $f_s = 1$ kHz.

Another requirement that must be fulfilled by the modified LS-SPWM is the even and efficient distribution of the zero states among the phase legs of the modules. A CSI module has three zero switching states, see Figure 2a–c. The modified LS-SPWM should be able to determine which zero state should be selected every time a zero state is required. In a previous study [15], the zero states distribution of the CSI is determined based on the extreme values of the three-phase reference currents. The same mechanism can be used here. Based on which phase has a positive or negative peak, the fundamental cycle can be divided into six intervals, as shown in Figure 5. For example, during interval I, phase A has a positive peak. Therefore, depending on m, S_{au} must be switched ON in most modules. During interval IV, on the other hand, phase A has a negative peak. Hence, S_{al} must be switched ON in most modules. During intervals I and IV, shortening leg A is the most efficient way to realize a zero state. Thus, during interval I, S_{au} is switched ON in all modules while S_{al} is switched ON in $M - i_{am}$ modules. Similarly, during interval IV, S_{al} is switched ON in all modules while S_{au} is switched ON in $M + i_{am}$ modules. Here, i_{am}, i_{bm}, and i_{cm} are the instantaneous values of the modulated three-phase currents, their values are between $-M$ and M. The same procedure can be repeated for the other four intervals where phases B and C have their positive and negative peaks. Table 1 summarizes the zero states distribution mechanism over one cycle. For each interval, the corresponding number of ON upper and lower switches of each phase of the MCSI is given.

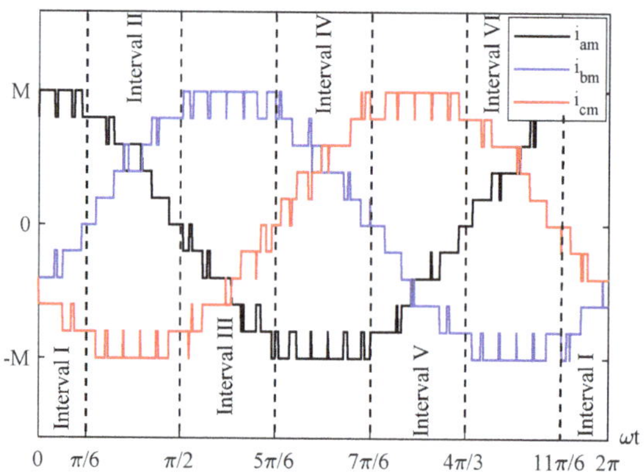

Figure 5. Intervals highlighting the extreme values of a set of three-phase currents.

Table 1. Zero state distribution in MCSI.

Interval	ON S_{au}	ON S_{al}	ON S_{bu}	ON S_{bl}	ON S_{cu}	ON S_{cl}
I	M	$M - i_{am}$	0	$-i_{bm}$	0	$-i_{cm}$
II	i_{am}	0	i_{bm}	0	$M + i_{cm}$	M
III	0	$-i_{am}$	M	$M - i_{bm}$	0	$-i_{cm}$
IV	$M + i_{am}$	M	i_{bm}	0	i_{cm}	0
V	0	$-i_{am}$	0	$-i_{bm}$	M	$M - i_{cm}$
VI	i_{am}	0	$M + i_{bm}$	M	i_{cm}	0

3. Current Balancing Algorithm

Table 1 provides the total number of ON switches for all the modules in the MCSI at any given instance. In odd intervals, the commands for the upper switches in each module are the same, the upper switch of a specified phase is turned ON in every module. On the other hand, the numbers of lower switches that should be turned ON in each phase of the MCSI are different. During Interval I, for example, Table 1 specifies how many modules should have their S_{al}, S_{bl}, and S_{cl} switched ON without mapping these commands to specific modules. During even intervals, the reverse is true. In this section, an optimized selection process referred to as CBA is presented. Based on the output line-to-neutral voltages v_{LN} and the sharing inductors' currents (i_{Lu} and i_{Ll} in each module), the CBA determines the optimum way to distribute the ON commands among the modules such that the average value of the inductors' currents remain close to each other.

Before introducing the CBA, it is important to understand the relationship between the output line-to-neutral voltages and the inductors currents. Consider the single module in Figure 6. Expressions for the voltages across the upper and lower inductors are given in Equation (10) and Equation (11), respectively. The voltage v_{cm} is the common-mode voltage, which has been discussed extensively in the literature [1]. Recall that only one of the upper switches (S_{au}, S_{bu}, and S_{cu}) can be switched ON, i.e., have a value of 1 while others have values of 0. The same constraint is applied to the lower switches. Equation (10) and Equation (11) show that the voltages across the upper and lower sharing inductors depends on which phase they are connected to. For example, assume $v_{aN} = 0.9$ p.u, $v_{bN} = -0.073$ p.u, and $v_{cN} = -0.827$ p.u. In this case, if the current in the upper sharing inductor is low, the best option is to connect it to phase C, by turning S_{cu} ON. Since Equation (10) has the highest possible value, it will result in the highest possible di_{Lu}/dt. If the current in the upper sharing inductor is high, on the other hand, connecting it to phase A would result in the lowest possible di_{Lu}/dt. The same conclusions are reversed for the lower sharing inductor, the line to neutral voltage in Equation (11) has a positive sign. Note that knowledge of v_{cm} is not required when determining the best option, because v_{cm} has the same effect across all phases. Therefore, to implement an optimized selection process, at every switching instance, the CBA starts by identifying what interval the MCSI is operating in. During odd intervals the CBA ranks the modules according to their lower sharing inductors' currents in a descending order. It then assigns the modules with low inductor's currents to phases that has the highest v_{LN}. During even intervals the CBA ranks the modules according to their upper sharing inductors' currents in a descending order; then assigns the modules with lowest inductor currents to phases that have the lowest v_{LN}. Hence, the average values of the upper and lower inductors' currents can be maintained around the same value, I_{dc}/M. The CBA is summarized in the flow chart presented in Figure 7.

$$v_{Lu} = \frac{V_{dc}}{2} - (S_{au}v_{aN} + S_{bu}v_{bN} + S_{cu}v_{cN}) - v_{cm} \tag{10}$$

$$v_{Ll} = \frac{V_{dc}}{2} + (S_{al}v_{aN} + S_{bl}v_{bN} + S_{cl}v_{cN}) + v_{cm} \tag{11}$$

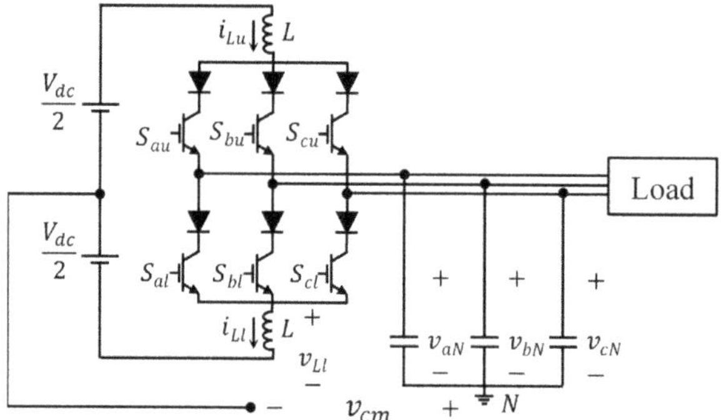

Figure 6. Single current source inverter (CSI) module.

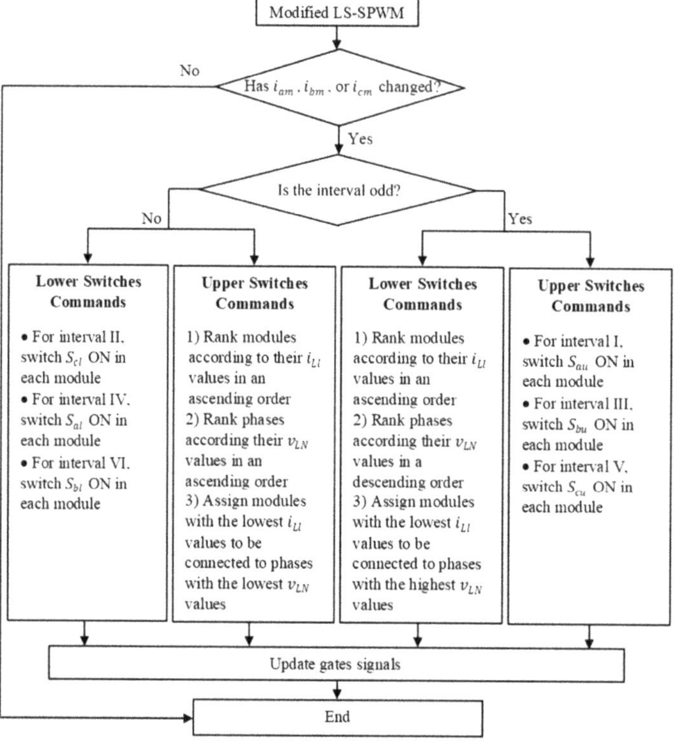

Figure 7. A flow chart summarizing the current balancing algorithm (CBA).

To illustrate the process of CBA through an example, assume the CBA was triggered by a PWM change in a five-level MCSI ($M = 2$) where the new values of i_{am}, i_{bm}, and i_{cm} are 2, −1, and −1, respectively. Furthermore, assume $v_{bN} = 0.5$ and $v_{cN} = −0.5$. Since the MCSI is operating in Interval I (i_{am} is at its positive peak), the CBA will switch S_{au} ON in both modules, switch S_{bl} ON in one module, and switch S_{cl} ON in the other module, i.e., one module will be assigned to S1 (see Figure 2d) while another module will be assigned to S2 (see Figure 2e). Here, the CBA must decide which module should receive the S_{bl} ON command and which one should receive the S_{cl} ON command. According to Equation (11), the value of v_{Ll} in a module is maximized when S_{bl} is switched ON (because $v_{bN} > v_{cN}$).

Therefore, the CBA will send the S_{bl} ON command to the module that has the lower i_{Ll} value while sending the S_{cl} ON command to the other module.

4. Experimental Results

A three-module, proof-of-concept prototype was built and tested to verify the operation of the MCSI. Unlike previous MCSI prototypes, having custom-designed inductors that are strictly identical was not required for proper operation of the converter [11]. In fact, the inductors used in the prototype are off-the-shelf 20 mH inductors with ± 5% tolerance [18]. The unidirectional switches used in the modules are realized by connecting a SiC diode in series with an IGBT. The modified LS-SPWM, CBA, and data acquisition were implemented in a dSPACE MicroLabBox. All the measured voltages and currents were sent to dSPACE ControlDesk and then plotted using Matlab. Figure 8 shows a picture of the experimental setup. The rest of the prototype's parameters are given in Table 2.

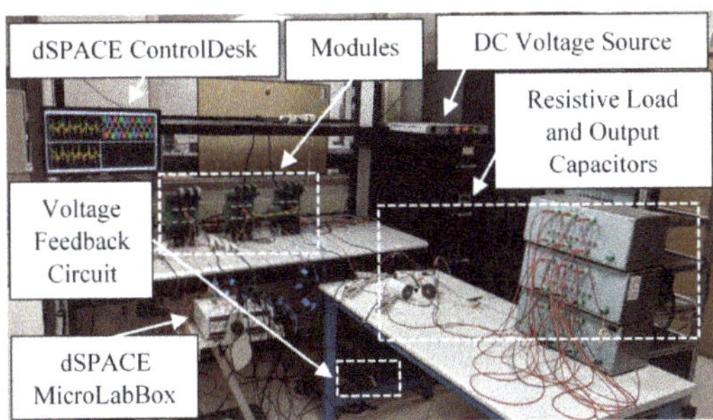

Figure 8. Experimental setup.

Table 2. Prototype parameters.

Parameter	Value
Number of Modules	$M = 3$
DC Voltage	30 V
Modulation Index	$m = 0.95$ (With 3rd harmonic injection)
AC Side Capacitors	$C = 100\ \mu\text{F}$ (Delta-Connected)
Load	Y-Connected 28.57 Ω Resistors
Sharing Inductor Parameters (5% Tolerance)	$L = 20\ \text{mH}$, $R_L = 0.558\ \Omega$
Switching Frequency	$f_s = 1389\ \text{Hz}$
Unidirectional Switch Parameters	IGBT: Infineon IRG7PH35UDPbF Diode: Cree C3D12065A
DSP	dSPACE MicroLabBox

Initially, only two modules were used to create a five-level MCSI. The operation of the five-level MCSI during startup is shown in Figure 9. The CBA's effectiveness is apparent in Figure 9b,c; the instantaneous values of the sharing inductors' currents remained close to each other during startup. Figure 10 shows the same quantities under steady-state conditions. The CBA's efforts can be observed in Figure 10b,c. When the output current of any phase is at its positive peak, referred to as odd intervals in Section 3, the upper sharing inductors experience similar di/dt, because they are connected to the same phase. On the other hand, the lower sharing inductors' currents are changing at various rates, which shows the CBA's effort to balance the inductors' currents. When the output current of any phase is at its negative peak, the same phenomenon is reversed, changes are more apparent in the upper inductor's currents.

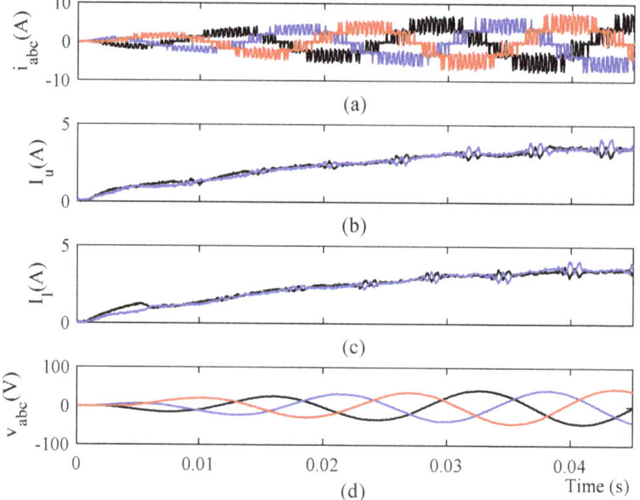

Figure 9. Operation of the five-level MCSI during startup: (**a**) output current, (**b**) upper inductors' currents, (**c**) lower inductors' currents and (**d**) output voltage.

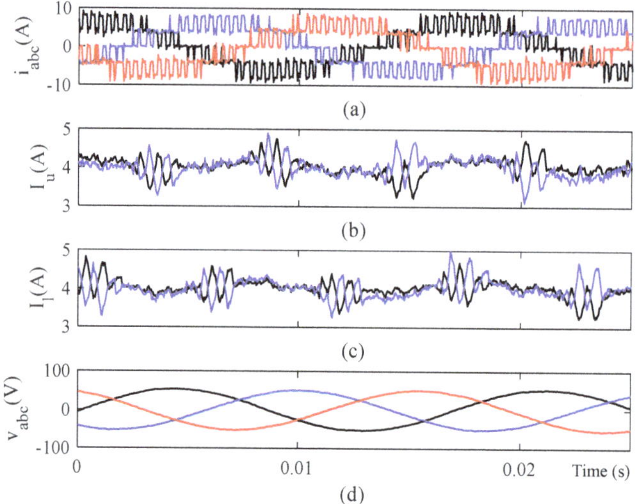

Figure 10. Steady-state operation of the five-level MCSI: (**a**) output current, (**b**) upper inductors' currents, (**c**) lower inductors' currents and (**d**) output voltage.

For further testing and verification, the third module of the prototype was activated to enable a seven-level operation. All the key waveforms of the seven-level MCSI during startup and steady-state operating conditions are presented in Figures 11 and 12, respectively. Further demonstration of the effectiveness of the CBA is shown in Figure 12b,c.

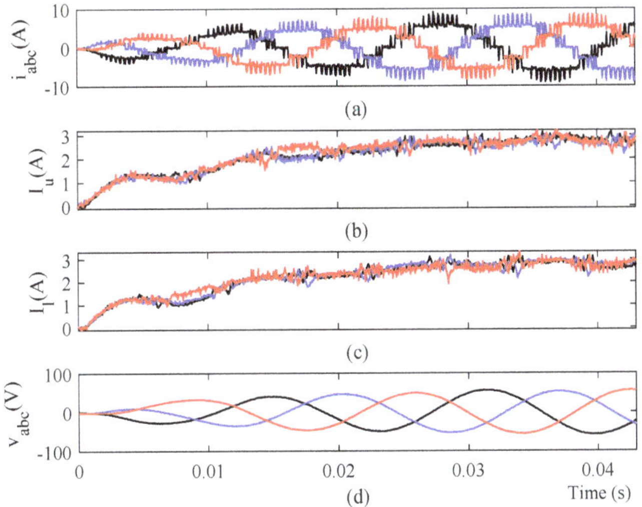

Figure 11. Operation of the seven-level MCSI during startup: (**a**) output current, (**b**) upper inductors' currents, (**c**) lower inductors' currents, and (**d**) output voltage.

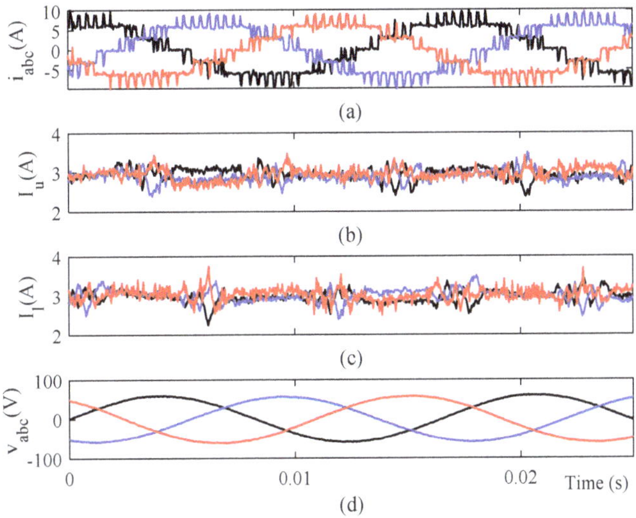

Figure 12. Steady-state operation of the seven-level MCSI: (**a**) output current, (**b**) upper inductors' currents, (**c**) lower inductors' currents and (**d**) output voltage.

5. Conclusions

A solution to the current balancing problem in voltage fed MCSIs was proposed in this paper. Unlike previous solutions [12,14], the CBA does not require any dedicated PI controllers; and it does not impose any alteration on the PWM carrier waveforms. Moreover, the CBA can be implemented in any MCSI, regardless of the number of levels. In addition, the modified LS-SPWM implementation in MCSIs was discussed in this paper. It was shown to have lower THD and di/dt compared to the PSC-SPWM. Although it was originally proposed for multilevel VSIs [16], it was never implemented in MCSIs. To validate the presented theoretical work, a proof-of-concept lab prototype was designed and tested. Experimental results for the five-level and seven-level cases were presented.

Author Contributions: Conceptualization, F.A. and M.G.S.; methodology, F.A. and M.G.S.; software, F.A.; validation, F.A.; formal analysis, F.A. and M.G.S.; investigation, F.A.; resources, F.A. and M.G.S.; data curation, F.A.; writing—original draft preparation, F.A.; writing—review and editing, F.A. and M.G.S.; visualization, F.A.; supervision, M.G.S.; project administration, M.G.S.; funding acquisition, F.A. and M.G.S. All authors have read and agreed to the published version of the manuscript.

Funding: This research received no external funding.

Conflicts of Interest: The authors declare no conflict of interest.

References

1. Wu, B.; Narimani, M. *High-Power Converters and AC Drives*; Wiley; IEEE Press: Hoboken, NJ, USA, 2017; ISBN 978-1-119-15606-2.

2. Moran, L.T.; Ziogas, P.D.; Joos, G. Analysis and design of a three-phase current source solid-state VAR compensator. *IEEE Trans. Ind. Appl.* **1989**, *25*, 356–365. [CrossRef]

3. Stretch, N.; Kazerani, M.; Shatshat, R.E. A Current-Sourced Converter-Based HVDC Light Transmission System. In Proceedings of the 2006 IEEE International Symposium on Industrial Electronics, Montreal, QC, Canada, 9–13 July 2006; Volume 3, pp. 2001–2006.

4. Gnanasambandam, K.; Rathore, A.K.; Edpuganti, A.; Srinivasan, D.; Rodriguez, J. Current-fed multilevel converters: An overview of circuit topologies, modulation techniques, and applications. *IEEE Trans. Power Electron.* **2016**, *32*, 3382–3401. [CrossRef]

5. Jin, J.X.; Tang, Y.J.; Xiao, X.Y.; Du, B.X.; Wang, Q.L.; Wang, J.H.; Wang, S.H.; Bi, Y.F.; Zhu, J.G. HTS power devices and systems: Principles, characteristics, performance, and efficiency. *IEEE Trans. Appl. Supercond.* **2016**, *26*, 1–26. [CrossRef]

6. Snitchler, G.; Gamble, B.; King, C.; Winn, P. 10 MW class superconductor wind turbine generators. *IEEE Trans. Appl. Supercond.* **2011**, *21*, 1089–1092. [CrossRef]

7. Chen, X.Y.; Jin, J.X.; Xin, Y.; Shu, B.; Tang, C.L.; Zhu, Y.P.; Sun, R.M. Integrated SMES technology for modern power system and future smart grid. *IEEE Trans. Appl. Supercond.* **2014**, *24*, 1–5. [CrossRef]

8. Bai, Z.; Zhang, Z. Conformation of multilevel current source converter topologies using the duality principle. *IEEE Trans. Power Electron.* **2008**, *23*, 2260–2267. [CrossRef]

9. Xiong, Y.; Chen, D.; Yang, X.; Hu, C.; Zhang, Z. Analysis and experimentation of a new three-phase multilevel current-source inverter. In Proceedings of the 2004 IEEE 35th Annual Power Electronics Specialists Conference (IEEE Cat. No. 04CH37551), Aachen, Germany, 20–25 June 2004; Volume 1, pp. 548–551.

10. Cossutta, P.; Aguirre, M.P.; Cao, A.; Raffo, S.; Valla, M.I. Single-stage fuel cell to grid interface with multilevel current-source inverters. *IEEE Trans. Ind. Electron.* **2015**, *62*, 5256–5264. [CrossRef]

11. Aguirre, M.P.; Calvino, L.; Valla, M.I. Multilevel current-source inverter with FPGA control. *IEEE Trans. Ind. Electron.* **2012**, *60*, 3–10. [CrossRef]

12. Aguirre, M.P.; Engelhardt, M.A.; Bracco, J.M.; Valla, M.I. Current balance control in a multilevel current source inverter. In Proceedings of the 2013 IEEE International Conference on Industrial Technology (ICIT), Cape Town, South Africa, 25–28 February 2013; pp. 1567–1572.

13. Bai, Z.; Zhang, Z. Digital control technique for multi-module current source converter. In Proceedings of the 2008 IEEE International Conference on Industrial Technology, Chengdu, China, 21–24 April 2008; pp. 1–5.

14. Cossutta, P.; Aguirre, M.P.; Engelhardt, M.A.; Valla, M.I. Control system to balance internal currents of a multilevel current-source inverter. *IEEE Trans. Ind. Electron.* **2017**, *65*, 2280–2288. [CrossRef]

15. Espinoza, J.; Joos, G. On-line generation of gating signals for current source converter topologies. In Proceedings of the ISIE'93-Budapest: IEEE International Symposium on Industrial Electronics Conference Proceedings, Budapest, Hungary, 1–3 June 1993; pp. 674–678.

16. Zhang, H.; Von Jouanne, A.; Dai, S.; Wallace, A.K.; Wang, F. Multilevel inverter modulation schemes to eliminate common-mode voltages. *IEEE Trans. Ind. Appl.* **2000**, *36*, 1645–1653.

17. Holmes, D.G.; Lipo, T.A. *Pulse width Modulation for Power Converters: Principles and Practice*; John Wiley & Sons: Hoboken, NJ, USA, 2003; Volume 18.

18. ERSE—Super Q. Available online: http://www.erseaudio.com/Products/SuperQCoilsAll/ESQ55-16-20000 (accessed on 31 December 2019).

Article

Accurate Deep Model for Electricity Consumption Forecasting Using Multi-Channel and Multi-Scale Feature Fusion CNN–LSTM

Xiaorui Shao, Chang-Soo Kim * and Palash Sontakke

Department of Information Systems, Pukyong National University, Busan 608737, Korea;
shaoxiaoruil@pukyong.ac.kr (X.S.); palashsntkk65@gmail.com (P.S.)
* Correspondence: cskim@pknu.ac.kr

Received: 13 March 2020; Accepted: 10 April 2020; Published: 12 April 2020

Abstract: Electricity consumption forecasting is a vital task for smart grid building regarding the supply and demand of electric power. Many pieces of research focused on the factors of weather, holidays, and temperatures for electricity forecasting that requires to collect those data by using kinds of sensors, which raises the cost of time and resources. Besides, most of the existing methods only focused on one or two types of forecasts, which cannot satisfy the actual needs of decision-making. This paper proposes a novel hybrid deep model for multiple forecasts by combining Convolutional Neural Networks (CNN) and Long-Short Term Memory (LSTM) algorithm without additional sensor data, and also considers the corresponding statistics. Different from the conventional stacked CNN–LSTM, in the proposed hybrid model, CNN and LSTM extracted features in parallel, which can obtain more robust features with less loss of original information. Chiefly, CNN extracts multi-scale robust features by various filters at three levels and wide convolution technology. LSTM extracts the features which think about the impact of different time-steps. The features extracted by CNN and LSTM are combined with six statistical components as comprehensive features. Therefore, comprehensive features are the fusion of multi-scale, multi-domain (time and statistic domain) and robust due to the utilization of wide convolution technology. We validate the effectiveness of the proposed method on three natural subsets associated with electricity consumption. The comparative study shows the state-of-the-art performance of the proposed hybrid deep model with good robustness for very short-term, short-term, medium-term, and long-term electricity consumption forecasting.

Keywords: smart grid; electricity forecasting; CNN–LSTM; very short-term forecasting (VSTF); short-term forecasting (STF); medium-term forecasting (MTF); long-term forecasting (LTF)

---

1. Introduction

Accurate, reliable, and timely electricity consumption information is the key to ensure a stable and efficient electricity supply. However, the electricity consumption in daily life usually fluctuates with time, region, season, temperature, and society. Even in the same city, electricity consumption in different areas may vary. Typically, the power company arranges fixed personnel to provide the electricity supply of the fixed place. Once there is a surge of local electricity consumption, the electricity supply of the area will be affected, thus affecting the healthy life. Forecasting actual future electricity consumption can make corresponding adjustments in time to avoid this situation. There are three types of forecasts according to the forecasting duration: short-term forecast (STF), medium-term forecasting (MTF), and long-term forecasting (LTF). Generally, STF focuses on the time range from 24 h to one week; MTF focuses on the time range from one week to one month, and LTF focuses on the time range longer than the other two types [1,2].

Different types of electric power forecasting have different purposes: The short-term electricity consumption forecasting supports the personnel and equipment arrangement of the next day. The medium-term electricity consumption forecasting gives decision support for the human resource allocation of the power company. The long-term electricity consumption forecasting is a significant decision basis from the macro perspective. To deal with an emergency such as line damage, natural disasters, and so on, very short-term (VST) power consumption forecasting is also essential. We defined very short-term electricity forecasting in this paper is hourly.

Different methods have been carried out for power forecasting, which mainly contains three categories: regression-based, time series-based, and machine learning-based methods [3]. The regression-based method can be divided into two sub-classes: Normal regression such as simple linear regression, lasso regression, ridge regression, and autoregression methods such as vector auto-regression (AR) and vector moving average (MA). Especially, Tang et al. applied a LASSO-based approach to forecast the current solar power generation by using the past 30 days of data and achieve better results than the support vector machine-based method [4]. Yu et al. applied an improved AR-based method for short-term hourly load forecasting, which was tested on two kinds of real-time hourly data sets [5]. Ordinary regression only considers the relationship of current variables and needs additional related data. However, the dependent variables are affected by the relevant variables of the current and past periods. The autoregressive model takes into account the impact of the current and past points, but it requires data that must be stationary. To overcome the disadvantages that occurred in the regression-based method, a time series-based method is presented for energy consumption forecasting. Autoregressive integrated moving average model (ARIMA) is one of the most excellent time series-based models. It not only considers the impact of the current and past periods but also can be used for non-stationary data. The ARIMA model can be symbolized as $ARIMA(p, d, q)$, where p is the parameter of lag p^{th} order autocorrelation, q is the parameter of lag q^{th} order partial autocorrelation, and d is the parameter for generating stationary time series. Usually, d ranges from 1 to 2; p, q range from 8 to 10 [3,6]. ARIMA has been employed for short-term power forecasting in [7,8]. Mitkov et al. [9] proved that ARIMA could be used for MTF and LTF for electricity forecasting.

The above regression-based and time series-based methods consider the relationship between the past and the current time is linear. However, most of the hidden relationships are nonlinear. The machine learning-based method can overcome this issue by using different nonlinear kernels such as support machine vectors (SVMs). Although some studies have successfully used SVM to predict energy consumption, there will be overfitting when data is broad [3,10]. Fortunately, the deep learning-based method can handle the overfitting problem very well with a good forecasting result. Recently, the convolutional neural network (CNN) [11], one of the mighty deep learning methods, has been widely applied for power forecasting due to its excellent feature extraction capacity. Li et al. [12] reshapes the data into two dimensions as an image and then applies CNN for short-term electrical load forecasting. A novel multi-scale CNN considering time-cognition was presented in [13] for multi-step short-term load forecasting. Suresh et al. developed a new sliding window algorithm to generate data to forecast solar PV using multi-head CNN in making STF and MTF [14]. Kim et al. applied CNN for VST photovoltaic power generation forecasting and compared it with the long short-term memory (LSTM) method, proving that the CNN-based method is better than LSTM for VSTF [15]. Another deep learning-based method LSTM was used for LTF and STF problems as it has long-term memory [16]. Ma et al. [17–20] employed LSTM for STF in the area of power. For LTF problems, Agrawal et al. presented a novel model by combining LSTM and recurrent neural network (RNN) to predict future five-year electricity loads [21]. An enhanced deep model was proposed in Han's work [22] for STF and MTF of electric load. The attention mechanism was combined into LSTM for short-term photovoltaic power forecasting in Zhou's work [23]. In order to overcome the shortcomings of a single model, some hybrid models are proposed for power forecasting, such asWang et al. [24] proposed ARIMA–LSTM for daily water level forecasting. It used LSTM to forecast the residuals through results and then utilized ARIMA to train the model with residuals. However, it is complex to build so many ARIMA models to get the residuals when the data size is massive. Another hybrid model, CNN–LSTM, is proposed in Kim's work [25] for minutely, hourly, daily, and weekly electricity energy consumption forecasting using multi-variables as input. Hu et al. [26] also applied CNN–LSTM for daily urban water demand forecasting using related meteorological data. However, collecting such correlated variables is hard and time-consuming in reality. Although Yan et al. proposed a hybrid of CNN–LSTM to predict power consumption by using raw time series, it only focused on VSTF (minutely) [27]. Moreover, Yan et al. [28] proposed a hybrid LSTM model, in which wavelet transform (WT) is applied to preprocess the raw univariate time series firstly. Later, stationary parts of transformation are selected for VSTF (minutely). However, there is a problem that occurred in Yan's work [28] is that we still need to select the stationary part by hand.

The limitations of current research for energy consumption forecasting are summarized as follows. On the one hand, most above methods only focused on one or two types of forecasts among VSTF, STF, MTF, and LTF. However, we need to master various types of future power consumption information to improve power supply efficiency and realize the smart grid. On the other hand, most existing methods refer to multi-variable regression, which requires collecting multiple related data. Motivated by this, we present a highly accurate deep model for various types of electricity forecasts by only using self-history data. We call this deep model multi-channels and scales CNN–LSTM (MCSCNN–LSTM). The proposed MCSCNN–LSTM employs dual channels as input to extract rich, robust feature representations from different domains of raw data. One channel is the raw sample, and the other is the information of statistics corresponding to the raw sample. We adopted the parallel structure of CNN–LSTM, which is different from conventional CNN–LSTM. At first, the CNN part in this structure extracts multi-scale and global features from the first channel using multi-scale and wide convolution technology. Then, the LSTM part guarantees to extract features that have a long-time dependency from the raw data. At last, combined with CNN, LSTM extracted features with statistics channels as comprehensive features to forecast the electricity consumption.

The biggest challenge is that the power consumption time series only has fewer time points rather than vibration signal, image, and video. It requires us to use CNN seriously due to the obtained data being relatively low dimensional. The strategy of this paper is to use a few pooling layers to reduce the loss of valuable information.

The main contributions of this paper are summarized as follows:

- To the best of our understanding, a few types of research focused on using one model for VSTF, STF, MTF, and LTF. This paper addresses this issue with MCSCNN–LSTM.
- The hybrid deep model MCSCNN–LSTM was designed, trained, and validated. The MCSCNN–LSTM obtains the highest performance compared to the current state-of-the-art methods.
- The proposed method can accurately forecast electricity consumption by inputting the self-history data without any additional data and any handcrafted feature selection operation. Therefore, it reduces the cost of data collection while simultaneously keeping high accuracy.
- The feature extraction capacity of each part has been analyzed.
- The excellent transfer learning and multi-step forecasting capacities of the proposed MCSCNN–LSTM have been proven.

The rest of the paper is arranged as follows. Section 2 formalizes our problem and gives the data generation method. Section 3 introduces the theoretical background of the proposed approach consisting of CNN, LSTM, and statistical components knowledge. Section 4 gives the proposed architecture for electricity forecasting. Each type of forecasting mission is defined in this section also. In Section 5, comparative experimental studies on three datasets are carried out. In Section 6, we discuss the proposed deep model. Section 7 presents the conclusions and feature work.

2. Problem Formulations

Our purpose is to forecast future power consumption using self-historical data. The self-historical data of power consumption could be expressed as a time series as follows:

$$T = (t_1, t_2, t_3, \ldots, t_i, \ldots, t_N)$$

(1)

where T contains N data points. Different types of forecasts have different elements in T. We defined four types of forecasts as shown in Table 1 and described follows.

- VSTF: Hourly forecasting, power consumption data of previous H hours are employed for next-hour power consumption forecasting.
- STF: Daily forecasting, applying power consumption data of previous D days to get the next day's power consumption.
- MTF: Weekly forecasting, using power consumption data of previous W weeks to forecast power consumption of the next week.
- LTF: Monthly forecasting, the power consumption data of previous M months are employed to get the next one month.

Table 1. Defined four types of forecasts.

Forecasts Types	Length of Input History Data	Outputs
VSTF (hourly)	Previous H hours	Next one hour
STF (daily)	Previous D days	Next one day
MTF (weekly)	Previous W weeks	Next one week
LTF (monthly)	Previous M months	Next one month

The time series T needs to reconstruct as Equation (2) to satisfy the input of the proposed deep model. The input matrix includes $N - L$ samples; the length of sample $x(t)$ is L. Different types of forecasts have different L, which corresponds to H, D, W, and M. The corresponding output is defined as Equation (3). Every output is the electricity consumption of the next duration.

$$Input = \begin{bmatrix} t_1 & t_2 & t_3 & \cdots & t_{L-1} & t_L \\ t_2 & t_3 & t_4 & \cdots & t_L & t_{L+1} \\ \vdots & \vdots & \vdots & \ddots & \vdots & \vdots \\ t_{N-L} & t_{N-L+1} & t_{N-L+2} & \cdots & t_{N-2} & t_{N-1} \end{bmatrix}$$

(2)

$$Output = \begin{bmatrix} t_{L+1} \\ t_{L+2} \\ \vdots \\ t_N \end{bmatrix}$$

(3)

3. Methods

3.1. CNN

CNN is a typical feedforward neural network. It virtually constructs various filters that can extract the characteristics of input data. Through these filters, the input data is convoluted and pooled, and the topology features hidden in the data are extracted step by step. With the deep entry of the network layer, the extracted features are abstracted. Therefore, the extracted features have translation, scaling, and rotation invariance. The sparse connection in CNN reduces the number of training parameters and speeds up the convergence; weight sharing effectively avoids algorithm overfitting; and downsampling makes full use of the features of the data and reduces the data dimension, optimizing the network structure [29,30]. CNN can deal with one-dimensional (1-D) signals and sequences, two-dimensional (2-D) images, and three-dimensional (3-D) videos. We apply CNN to extract features from 1-D sequences in this paper.

The essential components of CNN are convolutional operation and pooling operation. Through convolution operation, high-level local region feature representations are extracted with different filter kernels. The convolution process is described as follows:

$$x_j^l = f\left(\sum_{i \in M_j} x_i^{l-1} \times k_{ij}^l + B_j^l\right) \tag{4}$$

where x_j^l are the j feature maps of lth layer through convolution operation between $l - 1^{th}$'s output x_i^{l-1} and j filters k_{ij}^l, B_j^l is j bias of each feature map; i is in the range of j input values M_j. After convolution operation, x_j^l is processed with an activation function. The comprehensive result a_j^l is the input of the next layer. Rectified Linear Unit (ReLU) was widely applied to accelerate and converge the CNN, which enabled a nonlinear expression of input signals to enhance the representation ability. Which is formalized as follows:

$$a_j^l = \max\left(0, x_j^{l-1}\right) \tag{5}$$

Another key component of CNN is the pooling operation, which is employed to reduce the dimension of input data and ensure scale invariance. Thus, obtained features are more stable, especially when data is acquired from a noisy environment. There are three types of pooling operations: maximum, minimum, and average pooling operation. We give an example of utilizing maximum pooling, which is expressed as follows:

$$p_j^l = max\left(q_j^{l-1}(t)\right), \ t \in [(j-1)w, jw] \tag{6}$$

where p_j^l is the output of maximum value among $l - 1^{th}$ layer obtained feature maps $q_j^{l-1}(t)$, t is t^{th} output neurons at j^{th} layer in the network, w is the width of pooling size. Further details of CNNs can be found in LeCun's paper [11].

3.2. LSTM

The traditional feedforward neural networks only accept information from input nodes. They do not "remember" input to different time series [31]. Thus, it cannot extract the hidden features which have a long-time dependency from raw data. LSTM is proposed for overcoming this shortcoming as its long-term memory character [16]. It is a kind of special recurrent neural network (RNN). It implements memory function through gate structure in one cell as shown in Figure 1. The key point of the LSTM cell is the upper horizontal line, and it works like a conveyor belt; the information will not change during the transmission. It deletes old information or adds new information through three gate structures: forgot gate, input gate, and out gate. The output value of three gates and updated information are expressed using f_t, i_t, o_t, $\hat{C}_t$ as shown in the following formulas:

$$f_t = \sigma\left(w_f \cdot [h_{t-1}, x_t] + b_f\right) \tag{7}$$

$$i_t = \sigma\left(W_i \cdot [h_{t-1}, x_t] + b_i\right) \tag{8}$$

$$\hat{C}_t = \tan h\left(W_C \cdot [h_{t-1}, x_t] + b_C\right) \tag{9}$$

$$C_t = f_t * C_{t-1} + i_{t*}\hat{C}_t \tag{10}$$

$$o_t = \sigma\left(w_O \cdot [h_{t-1}, x_t] + b_O\right) \tag{11}$$

$$h_t = \sigma\left(w_O \cdot [h_{t-1}, x_t] + b_O\right) * \tan h(C_t) \tag{12}$$

where C_t represents the memory cell which integrates the old useful information $f_t * C_{t-1}$ and adds some new information $i_{t*}\hat{C}_t$. $W_{f,i,o}$ represents the weight and bias vectors of the abovementioned gates. σ is activation function *sigmoid*, h_{t-1} is the LSTM value of the previous time step, and x_t is input data.

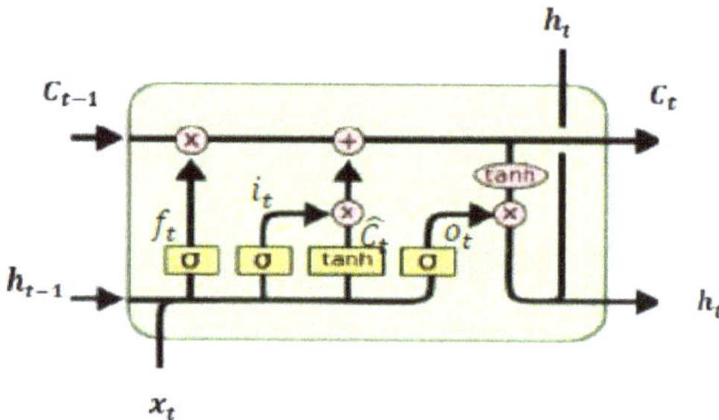

Figure 1. The structure of Long-Short-Term Memory (LSTM) cells.

3.3. Statistical Components

Statistics is a variable used to analyze and test data in statistical theory. It is the macro performance of data in the statistical domain. This paper creatively applied statistical components as one of the dual channels in the deep model to extract more features. The input matrix of raw time series we already defined as Equation (2). Each raw sample $x(t) \epsilon Input$ corresponds to six tuples named *Statistics*, which contains *mean*, *max*, *min*, standard deviation (*Sd*), skewness (*Skew*), and kurtosis (*Kurt*), which are defined as Equations (13)–(18).

$$mean(t) = \frac{1}{M} \sum_{t=1}^{M} x(t) \tag{13}$$

$$max(t) = max(x(t)) \tag{14}$$

$$min(t) = min(x(t)) \tag{15}$$

$$Sd(t) = \sqrt{\frac{1}{M} \sum_{t=1}^{M} (x(t) - mean(t))^2} \tag{16}$$

$$Skew(t) = E\left[\left(\frac{x(t) - mean(t)}{sd(t)}\right)^3\right] \tag{17}$$

$$Kurt(t) = E\left[\left(\frac{x(t) - mean(t)}{sd(t)}\right)^4\right] \tag{18}$$

4. Proposed Deep Model

We propose a deep model that has dual-channel inputs. One is raw data, and the other contains the six tuples of statistical components as we defined above. The overall architecture of the proposed deep model for electricity consumption forecasting can be seen from Figure 2 and a detailed configuration of the proposed deep model is shown in Table 2.

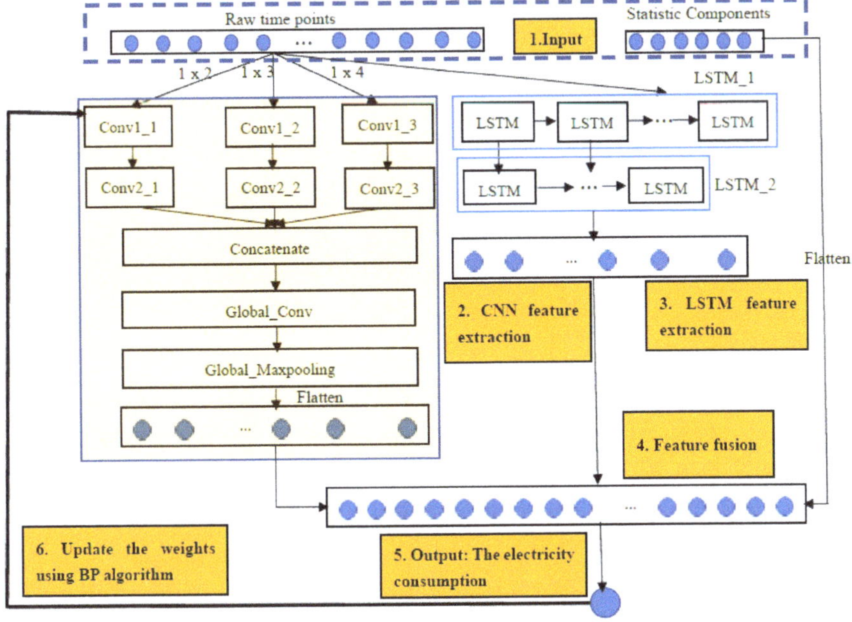

Figure 2. The architecture of the proposed multi-channels and scales convolutional neural networks (MCSCNN)–LSTM at three levels.

Table 2. Detailed configuration information of the proposed deep model.

Layer	Output Shape	Connected To	Parameters
Input1 (Raw)	(None, 24, 1)	–	0
Input2 (Statistic)	(None, 6, 1)	–	0
Conv1_1	(None, 12, 16)	Input1	48
Conv1_2	(None, 8, 16)	Input1	64
Conv1_3	(None, 6, 16)	Input1	80
Conv2_1	(None, 12, 16)	Conv1_1	528
Conv2_2	(None, 8, 16)	Conv1_2	528
Conv2_3	(None, 6, 16)	Conv1_3	528
Concatenate_1	(None, 26, 16)	Conv2_1, Conv2_2, Conv2_3	0
Static_Conv	(None, 11, 10)	Concatenate_1	2570
Global_Maxpooling	(None, 5, 10)	Static_Conv	0
Flatten_1	(None, 50)	Global_maxpooling	0
LSTM_1	(None, 24, 20)	Input2	1760
LSTM_2	(None, 10)	LSTM_1	1240
Flatten_2	(None, 6)	Input2	0
Concatenate_2	(None, 66)	LSTM_2, Flatten_1, Flatten_2	0
Dense (Output)	(None, 1)	Concatenate_2	67

Modifying the hyperparameters such as number and size of filter can improve the performance of the model. We defined the configuration information of MCSCNN–LSTM empirically. Here, we defined H, D, W, M as 24. The filter numbers of CNN decrease from 16 to 10 due to the shallow CNN layer being in charge of the detailed local feature extraction; the deeper CNN layer functions to capture abstract global feature representations. At the same time, LSTM is relatively time-consuming, so we defined proper output nodes in two LSTM layers as 20 and 10, respectively. From Figure 2, we can see six parts in our MCSCNN–LSTM: Input, CNN feature extraction, LSTM feature extraction, feature fusion, output, and weights updating. Every part is explained in detail as follows.

4.1. Input

The proposed deep model has double-channel inputs: raw sample $x(t)$ from the input matrix and six statistic components $Statistics(x(t))$, which can be written as:

$$In = \{(x(t), \ Statistics(x(t)))\} \tag{19}$$

Moreover, we transform the raw sample into one tensor with the shape of (24, 1), and six tuples *Statistics* into tensor with the shape of (6, 1) to satisfy the input requirements of the deep model. The reshaped tensor is defined in (20).

$$Tensor_{in} = \{Reshape(In)\} = \{(Reshape(x(t)), Reshape(Statistic(x(t))))\} \tag{20}$$

4.2. CNN Feature Extraction

Different from other CNNs, we adopted only one pooling layer to reduce the dimension of extracted features due to the data we used with lees dimensions, which is motivated by [31]. Firstly, CNN models the multi-scale local features from raw sample tensor $Reshape(x(t))$ at three-scale convolution operations—Conv1_1, Conv1_2, and Conv1_3—using different size kernels with shapes of 1×2, 1×3, 1×4. The convoluted results are activated by "ReLU", as defined in Equation (5). In order to obtain more robust features, we applied one more convolutional layer to extract the abstract feature representations again; they are Conv2_1, Conv2_2, and Conv2_3. At last, extracted multi-local features are processed by one wide convolution layer "Global_Conv" to obtain global representations. CNN extracted features are expressed as Equation (21) and then are flattened for the next step, where $CNN()$ is the process of this sub-section.

$$CNN_{features} = CNN(Reshape(x(t))) \tag{21}$$

4.3. LSTM Feature Extraction

Although CNN extracted rich feature representations, we doubt whether CNN can extract some critical hidden features having a long-time dependency. Based on this point, we employed LSTM to extract those features. Two-stacked LSTM layers are employed in this deep model, and every LSTM layer contains some LSTM cells as shown in Figure 1. The features LSTM extracted are expressed as Equation (22). $LSTM()$ is the process of this sub-section.

$$LSTM_{features} = LSTM(Reshape(x(t))) \tag{22}$$

4.4. Feature Fusion

Conventional CNN–LSTM is a stacked structure, in which CNN extracted features are processed by LSTM again. Different from conventional CNN–LSTM, this paper adopted a parallel pattern of CNN–LSTM to extract the features and then merged the features they extracted with flattening statistics components. Therefore, we obtained fusion features that are multi-scale and multi-domain (time and statistic domains), which are expressed as:

$$FUSION_{features} = Concatenate(CNN_{features}, \ LSTM_{features}, Flatten(Reshape(Statistic(x(t))))) \tag{23}$$

4.5. Output

After obtaining the rich, robust feature representations, the output is given by using one full connection layer between the output nodes and the fusion features, which can be defined as:

$$Output = W \cdot FUSION_{features} + B \tag{24}$$

where W is the weight matrix and B is the bias. This paper mainly focuses on power consumption forecasting of one duration unit (hourly, daily, weekly, and monthly). It is easy to extend the forecast of multi-duration units by setting the nodes of output; we will discuss this later.

4.6. Updating the Networks

Backpropagation [32] algorithm was employed for updating the weights of the hidden layer according to the loss function, and "Adam" [33] was selected as the optimizer for finding the convergence path. The loss function we applied in this deep model is Mean Squared Error (MSE) as shown in Equation (25), where y_i is ground truth electricity consumption and $\widetilde{y}_i$ is forecasting electricity consumption using the proposed hybrid deep model.

$$MSE = \frac{1}{N} \sum_{i=1}^{N} (y_i - \widetilde{y}_i)^2 \tag{25}$$

Electricity forecasting using the proposed model is formalized as Equation (26), where $MCSCNN\_LSTM()$ is our model, and $x(t)'$ is new history electricity consumption data points.

$$consumption = MCSCNN\_LSTM\big(x(t)', Statistic\big(x(t)'\big)\big) \tag{26}$$

5. Experiment Verification

In order to verify the effectiveness of the proposed deep model, we designed the following experiments using three datasets. The experiments are based on the operating system of Ubuntu 16.04.3, 64 bits with 23.4 GB RAM, and Intel (R) i7-700 CPU of processing speed 3.6 GHz. We used Keras to implement our proposed deep model.

5.1. Dataset Introduction

The data we adopted for validating the priority of the proposed method is from Pennsylvania-New Jersey-Maryland (PJM), which is a regional transmission organization in the USA. It is a part of the Eastern Interconnection grid operating an electric transmission system serving all or parts of some states. Different companies supply different regions. This paper applies three data sets from three companies: American Electric Power (AEP), Commonwealth Edison (COMED), and Dayton Power and Light Company (DAYTON). The raw data set of those is hourly consumption in megawatts (MW), and detailed information is described in Table 3. The data is available on the website of kaggle.com/robikscube/hourly-energy-consumption.

Table 3. Induction of data sets.

Dataset	Start Date	End Date	Length
AEP	2004-12-31 01:00:00	2018-01-02 00:00:00	121,273
COMED	2011-12-31 01:00:00	2018-01-02 00:00:00	66,497
DAYTON	2004-12-31 01:00:00	2018-01-02 00:00:00	121,275

The original data is utilized to validate the effectiveness of VSTF. For the other forecasting tasks, we use an overlapping sample algorithm to generate corresponding samples for each forecast, as shown in Algorithm 1. The stride of the algorithm we defined is one. Notably, we adopted the sample rate of 24 h, 168 h, and 720 h to generate each type of sample for STF, MTF, and LTF. One electricity consumption at different durations is given in Figure 3, in which different types fluctuate differently, respectively, VSTF and STF, which fluctuate frequently.

Algorithm 1: Overlapping sample algorithm

Input: Hourly electricity consumption historical time series *hourly*
Output: Daily, weekly, and monthly electricity consumption historical time series samples and labels.
Define the length of samples D, W, M as 24.
Step 1: Integrating the original data for different forecasts
       $daily < -sum(hourly, 24)$ #adopt the sample rate of 24 h for STF
       $weekly < -sum(hourly, 168)$ #adopt the sample rate of 168 h for MTF
       $monthly < -sum(hourly, 720)$ #adopt the sample rate of 720 h for LTF
Step 2: Generating the feature and label of each sample corresponding to the (2) and (3)
       For i in range $(length(daily/weekly/monthly))$: # different forecasts have different contents
            $daily_{features} < -daily[i : i + D]$
            $daily_{lables} < -daily[i + D + 1]$
            $weekly_{features} < -weekly[i : i + W]$
            $weekly_{labels} < -weekly[i + W + 1]$
            $monthly_{features} < -monthly[i : i + M]$
            $monthly_{labels} < -monthly[i + M + 1]$
    End for
Return $daily_{features}$, $daily_{labels}$, $weekly_{features}$, $weekly_{labels}$, $monthly_{features}$, $monthly_{labels}$

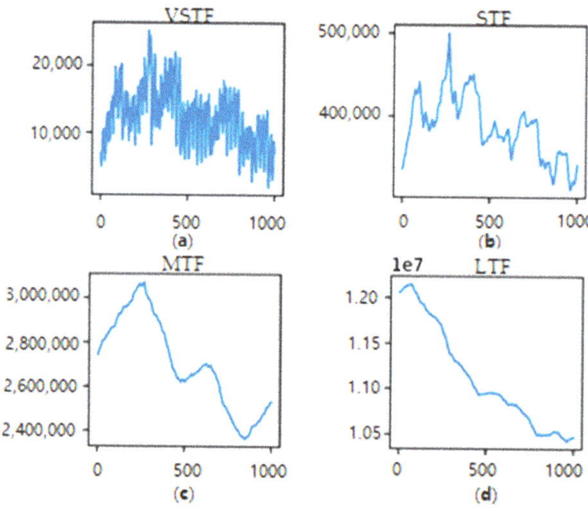

Figure 3. The electricity consumption at different durations. (**a**) Hourly electricity consumption for VSTF. (**b**) Daily electricity consumption for STF. (**c**) Weekly electricity consumption for MTF. (**d**) Monthly electricity consumption for LTF.

A description of each data for different forecasts is shown in Table 4. The first 80% of samples are utilized for training the model; the last 20% of samples are utilized to validate. Before starting the experiment, we adopted Equation (27) to normalize each data to work out the impact of different sizes of units, where x' is the normalized data point of time series T and x is the raw data sample.

$$x' = \frac{x - min(T)}{max(T) - min(T)} \tag{27}$$

Table 4. The description of each data set for different forecasts.

Forecasts	Dataset	Samples
VSTF (Hourly)	AEP	121,249
	COMED	66,473
	DAYTON	121,251
STF (Daily)	AEP	121,225
	COMED	66,449
	DAYTON	121,227
MTF (Weekly)	AEP	121,081
	COMED	66,305
	DAYTON	121,083
LTF (Monthly)	AEP	120,529
	COMED	65,753
	DAYTON	120,531

5.2. Evaluation Metrics

In order to fairly evaluate the effectiveness of the proposed MCSCNN–LSTM deep model, we adopted multiple evaluation metrics consisting of the root mean square error (RMSE), mean absolute error (MAE), and mean absolute percentage error (MAPE), as shown in Equations (28)–(30), where N is the number of testing samples, the *forecast* is the forecasted value, and *real* is the ground truth. RMSE evaluates the model by the standard deviation of the residuals between real values and forecasted values; MAE is the average vertical distance between ground truth values and forecasted values and is more robust to the larger errors than RMSE. However, when massive data are utilized for training and evaluating the model, the RMSE and MAE increase significantly and quickly. Therefore, MAPE is needed, which is the ratio between residuals and actual values.

$$RMSE = \sqrt{\frac{\sum_{n=1}^{N}(forecast_n - real_n)^2}{N}} \tag{28}$$

$$MAE = \frac{\sum_{n=1}^{N}|forecast_n - real_n|}{N} \tag{29}$$

$$MAPE = \frac{100\%}{N} \sum_{n=1}^{N}\left|\frac{forecast_n - real_n}{real_n}\right| \tag{30}$$

5.3. Performance Comparison with Other Excellent Methods

We compared our proposed method to other excellent deep learning-based methods: DNN- [34], NPCNN- [35], LSTM- [20], and CNN–LSTM-based [25] methods. The structure and configuration information of the above comparative methods are given in Table 5. Because the abovementioned methods employed other additional sensor data, we adopted the structure of them only. Conv1D is a convolutional layer with 1-D; Max1D is a max-pooling layer with 1D. We run 10 times of each deep learning-based method to overcome the impact of randomness, and every time runs at 50 epochs. Furthermore, we found the above NPCNN- [35], LSTM- [20], and CNN–LSTM-based [25] methods did not learn at some iterations. It means the loss does not decrease with the increase of training epochs. Instead, they keep one constant value from the first epoch. In summary, NPCNN, LSTM, and CNN–LSTM highly rely on initial processing. The results of this phenomenon are listed in Table 6, which gives the times of the above cases during 10-time training processes for each forecast. Notably, the term "None" means it always learns from raw data and is not sensitive to the random initial settings. For electricity consumption, we must avoid unpredicted and unexpected factors. However, NPCNN- [35], LSTM- [20], and CNN–LSTM-based [25] methods highly rely on initialization. The findings show only DNN [34], and the proposed method always learns so that they are considered as stable.

Table 5. The structure and configuration information of comparative methods.

Method	Structure# Layer (Neurons)	Activation Function
[34] DNN	Input-Dense(24)-Dense(10)-Flatten-Output	Sigmoid
[35] NPCNN	Input-Conv1D(5)-Max1D(2)-Flatten(Dense(1))-Dense(10)-Output	ReLu
[20] LSTM	Input–LSTM(20)–LSTM(20)-Output	ReLu
[25] CNN–LSTM	Input-Conv1D(64)-Max1D(2)-Conv1D(2)-Flatten(Max1D(2))–LSTM(64)-Dense(32)-Output	ReLu

Table 6. The non-training times of each deep learning-based method among 10 times.

Dataset	Method	VSTF	STF	MTF	LTF
	[34] DNN	None	None	None	None
	[35] NPCNN	2	3	2	2
AEP	[20] LSTM	5	4	5	2
	[25] CNN–LSTM	2	2	3	2
	Proposed	**None**	**None**	**None**	**None**
	[34] DNN	None	None	None	None
	[35] NPCNN	3	3	2	2
COMED	[20] LSTM	5	4	5	4
	[25] CNN–LSTM	2	3	4	2
	Proposed	**None**	**None**	**None**	**None**
	[34] DNN	None	None	None	None
	[35] NPCNN	2	2	3	2
DAYTON	[20] LSTM	4	4	2	3
	[25] CNN–LSTM	2	2	1	2
	Proposed	**None**	**None**	**None**	**None**

We compared the proposed method to the stable DNN [34] with averaged metrics of 10 times, and also compared averaged metrics of the proposed approach to the best results of three unstable methods: NPCNN [35], LSTM [20], and CNN–LSTM [25] in 10 times, as shown in Table 7 with RMSE, Table 8 with MAE, and Table 9 with MAPE. The findings reveal that our proposed method has absolute priority for different durations electricity forecasting at all evaluation metrics compared to DNN. Even when compared to the best results of the other three methods, the proposed MCSCNN–LSTM keeps the highest performance of all metrics for all data sets except for VSTF on the data set DAYTON with evaluation metric MAPE, and RMSE on data set AEP for STF. LSTM [20] performs a little better. In summary, the proposed MSCSNN–LSTM could forecast the electricity consumption of different durations accurately and stably.

The averaged improvements of MAPE on different data as shown in Figure 4. The results show that it improves a lot at all durations forecasts. Especially for STF, MTF, and LTF, which was beyond 50% compared to all the above methods. We select stable DNN as listed in Table 7 to compare the predicted results, as shown in Figure 5. The findings show both the proposed method and DNN [34] can predict the global trend of electricity consumption at VSTF, STF, and LTF. However, DNN cannot predict long-term electricity consumption. Moreover, the proposed method outperforms DNN; it can predict more detailed irregular trends for VSTF, STF, and LTF, respectively. We can see details from the marked deep-red box in VSTF and STF of Figure 5.

Table 7. The comparison results with RMSE.

Dataset	Method	RMSE (VSTF)	RMSE (STF)	RMSE (MTF)	RMSE(LTF)
AEP	[34] DNN	389.79	756.15	2864.03	15,387.72
	[35] NPCNN	476.38	1866.71	4220.96	40,393.06
	[20] LSTM	298.28	**124.19**	757.13	4876.94
	[25] CNN–LSTM	374.39	711.02	2408.09	20,060.97
	Proposed	**294.03**	424.14	**665.29**	**3385.70**
COMED	[34] DNN	310.69	765.16	3908.04	10,934.82
	[35] NPCNN	439.07	1090.17	6,274.38	14,900.91
	[20] LSTM	251.47	426.46	2925.53	30,407.07
	[25] CNN–LSTM	272.18	501.70	3082.33	4654.41
	Proposed	**240.51**	**377.74**	**520.02**	**3122.94**
DAYTON	[34] DNN	61.49	112.43	311.37	1299.99
	[35] NPCNN	71.16	183.90	390.88	1399.25
	[20] LSTM	43.84	142.49	107.87	444.95
	[25] CNN–LSTM	47.08	109.42	175.67	502.58
	Proposed	**43.68**	**65.68**	**95.84**	**270.40**

Table 8. The comparison results with MAE.

Dataset	Method	MAE (VSTF)	MAE (STF)	MAE (MTF)	MAE (LTF)
AEP	[34] DNN	246.41	583.44	2052.89	10,384.48
	[35] NPCNN	332.21	1682.52	2506.76	16,803.83
	[20] LSTM	198.67	995.27	613.45	3732.41
	[25] CNN–LSTM	248.65	508.70	1705.03	11,723.84
	Proposed	**180.94**	**250.15**	**494.04**	**2788.86**
COMED	[34] DNN	198.20	611.87	2951.25	8023.53
	[35] NPCNN	333.18	813.53	5427.71	11,953.35
	[20] LSTM	156.24	316.02	2831.15	30,082.99
	[25] CNN–LSTM	179.20	405.19	1181.74	3274.47
	Proposed	**142.60**	**244.50**	**345.84**	**2434.41**
DAYTON	[34] DNN	39.93	88.92	244.97	1145.00
	[35] NPCNN	49.81	135.24	312.21	1247.53
	[20] LSTM	29.10	116.53	613.45	372.93
	[25] CNN–LSTM	28.82	79.36	131.67	392.52
	Proposed	**27.12**	**38.68**	**70.04**	**212.64**

Table 9. The comparison results with MAPE.

Dataset	Method	MAPE (VSTF)	MAPE (STF)	MAPE (MTF)	MAPE (LTF)
AEP	[34] DNN	1.68	0.16	0.08	0.10
	[35] NPCNN	2.32	0.46	0.10	0.17
	[20] LSTM	1.65	0.27	0.03	**0.03**
	[25] CNN–LSTM	1.70	0.15	0.07	0.11
	Proposed	**1.23**	**0.06**	**0.02**	**0.03**
COMED	[34] DNN	1.79	0.23	0.16	0.10
	[35] NPCNN	3.02	0.30	0.29	0.15
	[20] LSTM	1.41	0.12	0.15	0.38
	[25] CNN–LSTM	1.68	0.15	0.07	0.04
	Proposed	**1.30**	**0.09**	**0.02**	**0.03**
DAYTON	[34] DNN	1.99	0.19	0.07	0.08
	[35] NPCNN	2.58	0.28	0.09	0.08
	[20] LSTM	**1.36**	0.23	0.03	0.03
	[25] CNN–LSTM	1.49	0.16	0.04	0.03
	Proposed	1.38	**0.08**	**0.02**	**0.01**

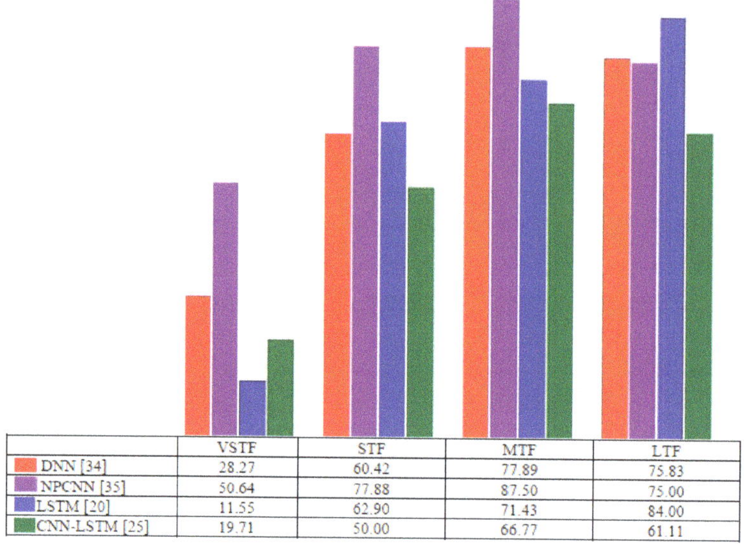

		VSTF	STF	MTF	LTF
■	DNN [34]	28.27	60.42	77.89	75.83
■	NPCNN [35]	50.64	77.88	87.50	75.00
■	LSTM [20]	11.55	62.90	71.43	84.00
■	CNN-LSTM [25]	19.71	50.00	66.77	61.11

Figure 4. The averaged MAPE improvement results for each duration forecast compared to other excellent deep learning-based methods on three data sets. Each data set contributes the same to the final outputs.

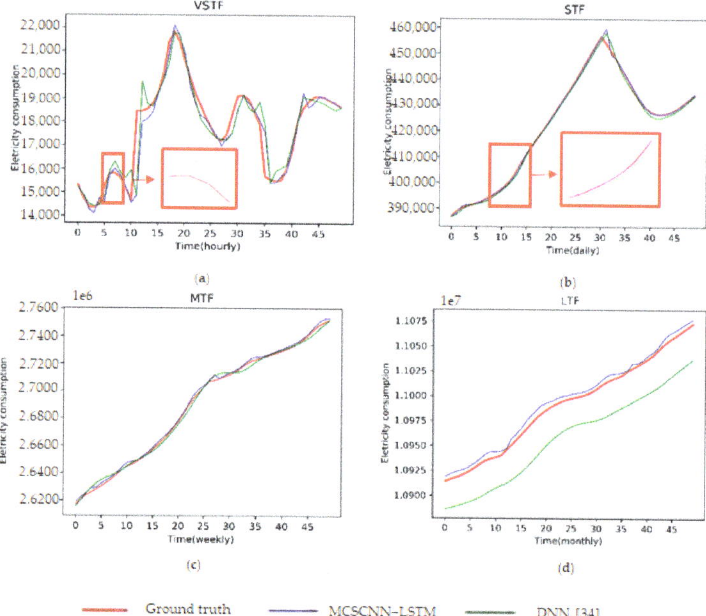

Figure 5. The comparative forecasting results using two different deep learning-based methods. The *x*-axis is the time stamp at different duration; the *y*-axis is electricity consumption. (**a**) VSTF (hourly) forecasting results. (**b**) STF (daily) forecasting results. (**c**) MTF (weekly) forecasting results. (**d**) LTF (monthly) forecasting results.

5.4. Feature Extraction Capacity of MCSCNN–LSTM

To better understand the feature extraction capacity of the proposed MCSCNN–LSTM, firstly, we compared it to some single models in MCSCNN–LSTM using the metric of MAPE: Multi-Scale CNN (MSCNN), Multi-Channel and Multi-Scale CNN (MCSCNN), and hybrid conventional stacked CNN–LSTM (SCNN–LSTM). The structure of LSTM is the same as [20], which is not stable. All configurations of those models are the same as MCSCNN–LSTM. The results are average of 10 times as shown in Table 10.

Table 10. The comparison results for validating the feature learning capacity using averaged MAPE.

Dataset	Method	VSTF	STF	MTF	LTF
AEP	MSCNN	1.91	1.17	0.85	1.26
	MCSCNN	2.28	0.93	0.79	1.19
	SCNN–LSTM	1.84	0.64	0.57	0.75
	Proposed	**1.23**	**0.06**	**0.02**	**0.03**
COMED	MSCNN	2.36	1.02	0.79	0.86
	MCSCNN	2.31	0.87	0.89	0.58
	SCNN–LSTM	1.93	0.80	1.08	0.43
	Proposed	**1.30**	**0.09**	**0.02**	**0.03**
DAYTON	MSCNN	2.28	1.14	0.77	0.91
	MCSCNN	2.61	0.81	0.70	0.83
	SCNN–LSTM	2.08	0.63	0.45	0.45
	Proposed	**1.38**	**0.08**	**0.02**	**0.01**

Comparing the MSCNN with the NPCNN [35], we find that the proposed MSCNN is more stable than general CNN-based methods. The results indicate MSCNN with single input cannot extract satisfactory feature representations for different forecasts of electricity consumption by comparing MSCNN with MCSCNN, especially for STF, MTF, and LTF. By comparing SCNN–LSTM to LSTM, we can see the MCSCNN can extract elegant, robust features to avoid instability and improve performance. We computed the averaged improvement of MAPE on three data sets for different duration forecasts, as shown in Figure 6. MSCNN is selected as the baseline. The findings reveal that the proposed method promotes a lot for all kinds of forecasts. Attentively, the results show only the performance of the MCSCNN on data AEP for VSTF decreased. Furthermore, the level of feature extraction capacity is ranked as proposed: > SCNN–LSTM > MCSCNN > MSCNN.

Secondly, we have analyzed the inside features to confirm the productive feature extraction capacity of the proposed deep model. The visualization results using one VSTF sample shown in Figure 7a,b intimate double-channel inputs: one raw data sample and corresponding statistics components by using the normalized data sample of AEP. Figure 7c is a CNN-learned feature map through the raw data sample. Figure 7d is the feature map of LSTM and we marked it with a red box at the comprehensive feature map Figure 7f, and Figure 7e is a statistic feature map that is marked with the yellow box in Figure 7f. The unmarked part in Figure 7f is a CNN-learned feature map. Figure 7f is a comprehensive feature map. The findings reveal that CNN can learn multi-scale robust global features with less noise because it almost has no changes around 0. The feature map of LSTM ranges from −0.100 to 0.075, and the statistic feature map ranges from 0.00 to 1.75, which indicates statistic components are more useful to extract detailed patterns than LSTM. The comprehensive feature maps combined robust multi-scale global features and detailed features of different domains.

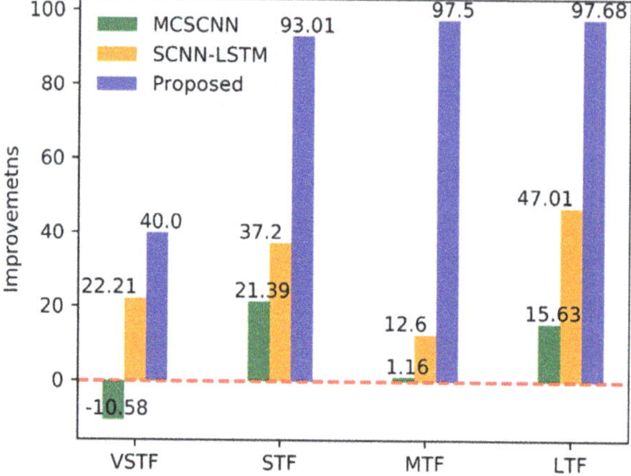

Figure 6. The averaged improvement of the proposed deep model for each duration forecast based on MSCNN on three data sets.

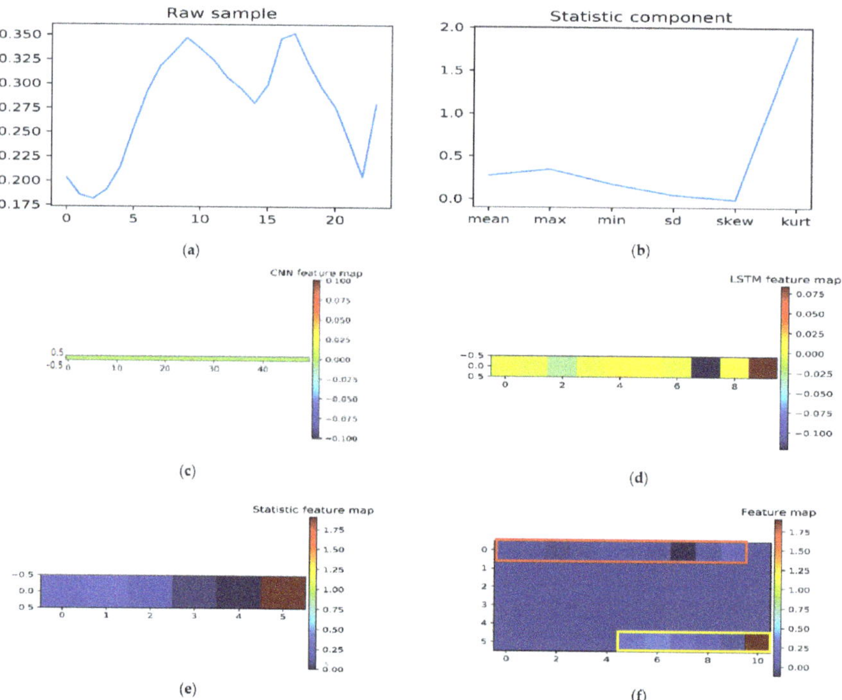

Figure 7. Feature maps visualization. Each part of the proposed deep model extracted different features. (**a**) The raw sample channel. (**b**) The statistic components channel. (**c**) CNN-learned feature map, which almost has no changes around 0. (**d**) LSTM-learned feature map, which ranges from –0.10 to 0.075. (**e**) Statistic components feature map of a reshaped tensor. The raw statistic components channel was reshaped into [1,6], which ranges from 0.00 to 1.75. (**f**) Reshaped comprehensive feature map. The shape of the obtained feature is 1 by 66, and we reshaped it into 11 by 6 to clearly see and analyze.

5.5. Transfer Learning Capacity Test

We have validated the transfer learning capacity of the proposed method to satisfy the needs in the real-life. For instance, one new company wants to predict electricity consumption, but they do not have enough historical data to train the model. It requires the model with an excellent transfer learning capacity. The following experiments are designed to test the transfer learning capacity of the proposed method, as described in Table 11. We adopted the training part of AEP as the training set to train the model for VSTF, the training part of COMED to train the model for STF, and the training part of DAYTON to train the model for MTF and LTF. The testing part of others is utilized to test.

Table 11. Experiment design for transfer learning capacity test of the proposed deep model.

Forecasts	Training Sets	Testing Sets
VSTF	AEP	COMED, DAYTON
STF	COMED	AEP, DAYTON
MTF	DAYTON	AEP, COMED
LTF	DAYTON	AEP, COMED

The DNN [34] and the proposed MCSCNN–LSTM applied the same data to train and test are considered as comparative experiments to validate the transfer learning capacity. For example, we trained DNN and MCSCNN–LSTM with the training part of COMED, DAYTON, and tested on the testing part of the same data set for VSTF. The results as shown in Figure 8; the x-axis is the testing part of each data set. The results indicate the proposed method has a functional transfer learning capacity, which outperforms DNN [34] for all kinds of forecasts, and a little lower than the proposed method using the same data to train and test the model. We performed a t-test to quantify this difference. The results of the p-value are shown in Table 12. If a p-value is higher than 0.05, it means there is no significant difference. The results show there was no significant difference when we utilized different companies' data for training the model. Moreover, even though DNN [34] employed the same source data to train and test model, its performance is worse than "transfer". Notably, there is a significant improvement for the VSTF of electricity consumption compared to DNN. In summary, Figure 8 and Table 12 confirmed that the proposed method has an excellent transfer learning capacity against noisy data.

Table 12. The p-value of significance test using t-test.

Forecasts	Transfer vs. DNN [34]	Transfer vs. Proposed
VSTF	0.0380	0.4650
STF	0.4800	0.3600
MTF	0.3120	0.1840
LTF	0.1300	0.4230

5.6. Multi-Step Forecasting Capacity Test

Accurate one-step forecasting enables decision-makers to create proper policies and measures of the power supply before one duration. Multi-step forecasting can provide multi-step future consumption information in advance. To validate the multi-step forecasting capacity of the proposed method for multiple forecasts, we designed a five-step forecasting experiment and compared it to [27], which tests the multi-step forecasting capacity of their model for VSTF. The input of the proposed MCSCNN–LSTM for multi-step forecasting in Figure 2 has the same shape with one-step forecasting, we only need to change the output into one vector with five elements regarding the five future consumption data points of each forecast. The comparative results using averaged RMSE, MAE, and MAPE of 10 times are shown in Table 13. AEP is adopted for VSTF and LTF; COMED and DAYTON are adopted for STF and MTF. The results indicate the proposed MCSCNN–LSTM performs very well for STF, MTF, and LTF. Especially, the performance of MTF and LTF has increased tenfold compared to the method of [27] using RMSE, MAE, and MAPE. Only the VSTF is a little worse than [27], but they still are the same level. We also give one sample of five-step forecasting of different forecasts as shown in Figure 9. We can see the proposed MCSCNN–LSTM accurately predicts all types of trends from the raw data and it outperforms [27] CNN–LSTM in terms of handing details. Notably, the proposed method has an absolute advantage in terms of STF, MTF, and LTF.

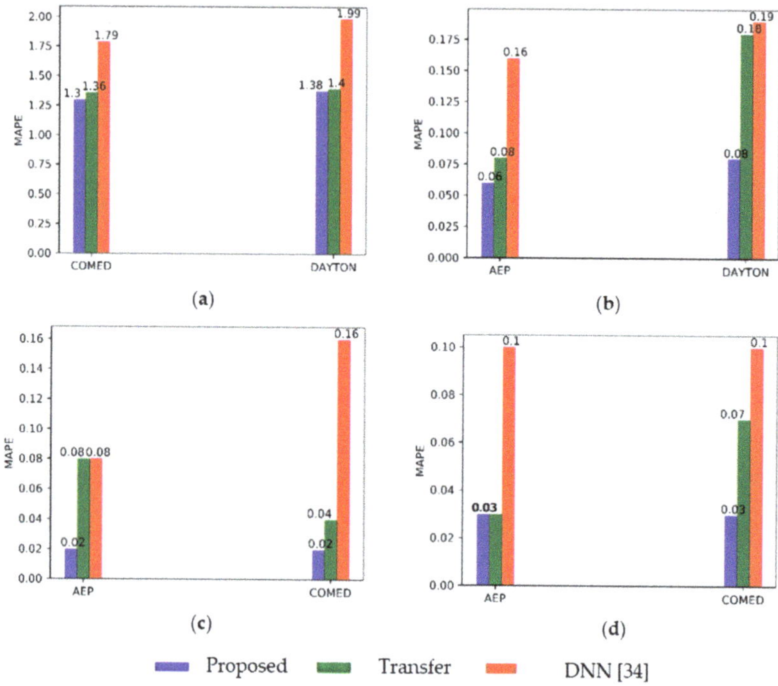

Figure 8. The average MAPE of different methods to validate the transfer learning capacity of the proposed MCSCNN–LSTM for different forecasts. (**a**) Transfer learning capacity test for VSTF. (**b**) Transfer learning capacity test for STF. (**c**) Transfer learning capacity test for MTF. (**d**) Transfer learning capacity test for LTF.

Table 13. The multi-step forecasting capacity test results.

Metrics	RMSE		MAE		MAPE	
Forecasts (Data)	Proposed	CNN–LSTM [27]	Proposed	CNN–LSTM [27]	Proposed	CNN–LSTM [27]
VSTF(AEP)	508.33	477.08	354.52	324.73	2.43	2.22
STF(COMED)	1.8311×10^3	1.8320×10^3	1.3592×10^3	1.3612×10^3	0.50	0.51
MTF(DAYTON)	4.6342×10^2	1.0418×10^3	3.0828×10^2	8.2458×10^2	0.11	0.23
LTF(AEP)	1.1251×10^4	3.0284×10^4	6.9459×10^3	2.3523×10^4	0.07	0.22

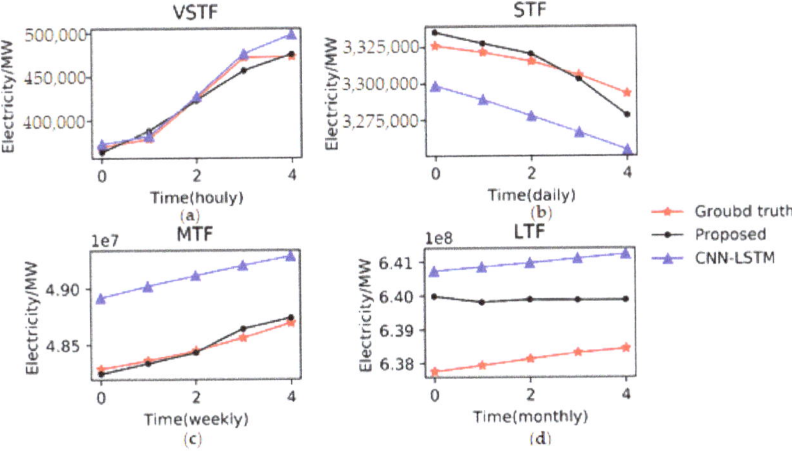

Figure 9. A comparison of the results of the five-step forecasting using the proposed method and CNN–LSTM [27]. The results indicate the proposed method has an absolute advantage in terms of STF, MTF, and LTF. For VSTF, CNN–LSTM performs a little better than proposed MCSCNN–LSTM. (**a**) Five-step electricity forecasting results for VSTF. (**b**) Five-step electricity forecasting results for STF. (**c**) Five-step electricity forecasting results for MTF. (**d**) Five-step electricity forecasting results for LTF.

6. Discussion

We have proposed a novel MCSCNN–LSTM that models different domain and multi-scale feature patterns to forecast the electricity consumption at different durations. The difficulty of feature extraction for different durations forecasts using one model, and different duration electricity consumptions have different patterns of the trend as shown in Figure 3, which requires a model with excellent feature extraction capacity with good robustness. Besides, collecting related data such as weather, the temperature is costly and time-consumption. Therefore, we developed MCSCNN–LSTM to extract multi-scale and multi-domain features by only inputting the electricity history data, as shown in Figure 2. We connected CNN and LSTM parallelly with dual inputs, Table 10 and Figure 6 shows it is more effective than conventional stacked CNN–LSTM.

We compared our proposed deep model with other excellent deep models in Table 6, which indicates our proposed model is stable. Furthermore, Tables 7–9 and Figure 4 show that we have improved the performance compared to the stable DNN [34] and the best results of NPCNN [35], LSTM [20], and CNN–LSTM [25]. Primarily, it has improved a lot for STF, MTF, and LTF. Figure 5 explained that the proposed method could predict the detailed irregular patterns of electricity consumption.

As can be seen from Table 10, we have analyzed the feature capacity of each part of the proposed MCSCNN–LSTM by comparing the averaged MAPE. In addition, we computed the averaged improvement ratio of the proposed MCSCNN–LSTM by using MCSCNN as the benchmark in Figure 6. It proved that each part of our proposed model has excellent feature extraction capacity. Moreover, we have analyzed the inside feature map of MCSCNN–LSTM as shown in Figure 7, it shows the CNN part of MCSCNN–LSTM can extract multi-scale robust global features, and statistic components are more effective in extracting detailed patterns than LSTM.

As shown in Figure 8, we have designed comparative experiments on three data sets to validate the transfer learning capacity of the proposed MCSCNN–LSTM. The findings from Figure 8 have proven that our proposed deep model has excellent transfer learning skills. In order to quantify the transfer learning capacity, we compared the p-value of the t-test with none-transfer learning methods in Table 12. Moreover, we have confirmed the proposed method could accurately forecast multi-step electricity consumption in advance in Table 13 and Figure 9, the results from Table 13 and Figure 9 indicate our proposed method outperforms CNN–LSTM which was developed in [27].

7. Conclusions

In conclusion, we proposed a novel MCSCNN–LSTM to forecast the electricity consumption at different durations accurately and robustly only by using the self-history data. The comparative analysis has shown that the proposed hybrid deep model MCSCNN–LSTM reaches state-of-the-art performance. The proposed model is compared to other excellent deep learning-based methods to confirm the efficiency and robustness. We run ten times for each model on three data sets to evaluate fairly. The results indicate that our proposed deep model is not sensitive to the initial settings and stable. We compare the forecasted results with other methods to prove that the proposed method can extract more detailed patterns. We also confirmed the necessity of each part in the proposed deep model by comparing the MAPE of each part for electricity forecasting at different durations. We proved that the parallel structure of CNN–LSTM is more potent than conventional stacked CNN–LSTM. We also analyzed the internal feature maps to confirm the feature extraction capacity of each part, and the results show CNN can extract global features; LSTM, and statistic components are in charge of detailed pattern extraction. Some individual experimental cases are designed to validate their excellent transfer learning capacity. We confirmed the proposed MCSCNN–LSTM has excellent multi-step forecasting capacity for STF, MTF, and LTF, respectively. The proposed MCSCNN–LSTM can accurately and stably predict the irregular patterns of electricity consumption at different durations by only using self-history data and have a good transfer learning capacity, which can be easy to extend to other forecasting tasks.

In this paper, we designed the networks empirically. Setting proper hyperparameters can effectively improve forecasting performance. In the feature, we will use deep reinforcement learning to automatically build the model and choose the better hyperparameters of MCSCNN–LSTM for electricity consumption forecasting.

Author Contributions: Designed algorithm, X.S. and C.-S.K.; Performed simulations, X.S.; Preprocessed and analyzed the data X.S. and P.S.; Wrote the paper, X.S.; Provide ideas to improve the performance of algorithm, X.S. and C.-S.K. All authors have read and agreed to the published version of the manuscript.

Funding: This work was supported by the Technology Innovation Program (20004205, The development of smart collaboration manufacturing innovation service platform in textile industry by producer-buyer B2B connection funded By the Ministry of Trade, Industry & Energy (MOTIE, Korea)).

Conflicts of Interest: The authors declare no conflicts of interest.

References

1. Ahmad, A.S.; Hassan, M.Y.; Abdullah, M.P.; Rahman, H.A.; Hussin, F.; Abdullah, H.; Saidur, R. A review on applications of ANN and SVM for building electrical energy consumption forecasting. *Renew. Sustain. Energy Rev.* **2014**, *33*, 102–109. [CrossRef]
2. Ayoub, N.; Musharavati, F.; Pokharel, S.; Gabbar, H.A. ANN Model for Energy Demand and Supply Forecasting in a Hybrid Energy Supply System. In Proceedings of the 2018 IEEE International Conference on Smart Energy Grid Engineering (SEGE), Oshawa, ON, Canada, 12–15 August 2018.
3. Weng, Y.; Wang, X.; Hua, J.; Wang, H.; Kang, M.; Wang, F.Y. Forecasting Horticultural Products Price Using ARIMA Model and Neural Network Based on a Large-Scale Data Set Collected by Web Crawler. *IEEE Trans. Comput. Soc. Syst.* **2019**, *6*, 547–553. [CrossRef]
4. Tang, N.; Mao, S.; Wang, Y.; Nelms, R.M. Solar Power Generation Forecasting With a LASSO-Based Approach. *IEEE Internet Things J.* **2018**, *5*, 1090–1099. [CrossRef]
5. Yu, J.; Lee, H.; Jeong, Y.; Kim, S. Short-term hourly load forecasting using PSO-based AR model. In Proceedings of the 2014 Joint 7th International Conference on Soft Computing and Intelligent Systems (SCIS) and 15th International Symposium on Advanced Intelligent Systems (ISIS), Kitakyushu, Japan, 3–6 December 2014; pp. 1449–1453.

6. Wu, H.; Wu, H.; Zhu, M.; Chen, W.; Chen, W. A new method of large-scale short-term forecasting of agricultural commodity prices: Illustrated by the case of agricultural markets in Beijing. *J. Big Data* **2017**, *4*, 1–22. [CrossRef]

7. Radziukynas, V.; Klementavicius, A. Short-term wind speed forecasting with ARIMA model. In Proceedings of the 2014 55th International Scientific Conference on Power and Electrical Engineering of Riga Technical University (RTUCON), Riga, Latvia, 14 October 2014; pp. 145–149.

8. He, H.; Liu, T.; Chen, R.; Xiao, Y.; Yang, J. High frequency short-term demand forecasting model for distribution power grid based on ARIMA. In Proceedings of the 2012 IEEE International Conference on Computer Science and Automation Engineering (CSAE), Zhangjiajie, China, 25–27 May 2012; pp. 293–297.

9. Mitkov, A.; Noorzad, N.; Gabrovska-Evstatieva, K.; Mihailov, N. Forecasting the energy consumption in Afghanistan with the ARIMA model. In Proceedings of the 2019 16th Conference on Electrical Machines, Drives and Power Systems (ELMA), Varna, Bulgaria, 6–8 June 2019; pp. 1–4.

10. Jha, G.K.; Sinha, K. Agricultural Price Forecasting Using Neural Network Model: An Innovative Information Delivery System. *Agric. Econ. Res. Rev.* **2013**, *26*, 229–239. [CrossRef]

11. Lecun, Y.; Bengio, Y.; Hinton, G. Deep learning. *Nature* **2015**, *521*, 436–444. [CrossRef]

12. Li, L.; Ota, K.; Dong, M. Everything is image: CNN-based short-term electrical load forecasting for smart grid. In Proceedings of the 2017 14th International Symposium on Pervasive Systems, Algorithms and Networks & 2017 11th International Conference on Frontier of Computer Science and Technology & 2017 Third International Symposium of Creative Computing (ISPAN-FCST-ISCC), Exeter, UK, 21–23 June 2017; pp. 344–351.

13. Deng, Z.; Wang, B.; Xu, Y.; Xu, T.; Liu, C.; Zhu, Z. Multi-scale convolutional neural network with time-cognition for multi-step short-Term load forecasting. *IEEE Access* **2019**, *7*, 88058–88071. [CrossRef]

14. Suresh, V.; Janik, P.; Rezmer, J.; Leonowicz, Z. Forecasting solar PV output using convolutional neural networks with a sliding window algorithm. *Energies* **2020**, *13*, 723. [CrossRef]

15. Kim, D.; Hwang, S.W.; Kim, J. Very Short-Term Photovoltaic Power Generation Forecasting with Convolutional Neural Networks. In Proceedings of the 9th International Conference on Information and Communication Technology Convergence (ICTC), Jeju, Koera, 17–18 October 2018; pp. 1310–1312.

16. Hochreiter, S.; Urgen Schmidhuber, J. Long Shortterm Memory. *Neural Comput.* **1997**, *9*, 17351780. [CrossRef]

17. Ma, Y.; Zhang, Q.; Ding, J.; Wang, Q.; Ma, J. Short Term Load Forecasting Based on iForest-LSTM. In Proceedings of the 2019 14th IEEE Conference on Industrial Electronics and Applications (ICIEA), Xi'an, China, 19–21 June 2019; pp. 2278–2282.

18. Xiuyun, G.; Ying, W.; Yang, G.; Chengzhi, S.; Wen, X.; Yimiao, Y. Short-term Load Forecasting Model ofGRU Network Based on Deep Learning Framework. In Proceedings of the 2018 2nd IEEE Conference on Energy Internet and Energy System Integration (EI2), Beijing, China, 20–22 October 2018; pp. 1–4.

19. Yu, Y.; Cao, J.; Zhu, J. An LSTM Short-Term Solar Irradiance Forecasting under Complicated Weather Conditions. *IEEE Access* **2019**, *7*, 145651–145666. [CrossRef]

20. Kong, W.; Dong, Z.Y.; Jia, Y.; Hill, D.J.; Xu, Y.; Zhang, Y. Short-Term Residential Load Forecasting Based on LSTM Recurrent Neural Network. *IEEE Trans. Smart Grid* **2019**, *10*, 841–851. [CrossRef]

21. Agrawal, R.K.; Muchahary, F.; Tripathi, M.M. Long term load forecasting with hourly predictions based on long-short-term-memory networks. In Proceedings of the 2018 IEEE Texas Power and Energy Conference (TPEC), College Station, TX, USA, 8–9 February 2018; pp. 1–6.

22. Han, L.; Peng, Y.; Li, Y.; Yong, B.; Zhou, Q.; Shu, L. Enhanced deep networks for short-term and medium-term load forecasting. *IEEE Access* **2019**, *7*, 4045–4055. [CrossRef]

23. Zhou, H.; Zhang, Y.; Yang, L.; Liu, Q.; Yan, K.; Du, Y. Short-Term photovoltaic power forecasting based on long short term memory neural network and attention mechanism. *IEEE Access* **2019**, *7*, 78063–78074. [CrossRef]

24. Wang, Z.; Lou, Y. Hydrological time series forecast model based on wavelet de-noising and ARIMA-LSTM. In Proceedings of the 2019 IEEE 3rd Information Technology, Networking, Electronic and Automation Control Conference (ITNEC 2019), Chengdu, China, 15–17 March 2019; pp. 1697–1701.

25. Kim, T.Y.; Cho, S.B. Predicting residential energy consumption using CNN-LSTM neural networks. *Energy* **2019**, *182*, 72–81. [CrossRef]

26. Hu, P.; Tong, J.; Wang, J.; Yang, Y.; Oliveira Turci, L. De A hybrid model based on CNN and Bi-LSTM for urban water demand prediction. In Proceedings of the 2019 IEEE Congress on Evolutionary Computation (CEC), Wellington, New Zealand, 10–13 June 2019; pp. 1088–1094.
27. Yan, K.; Wang, X.; Du, Y.; Jin, N.; Huang, H.; Zhou, H. Multi-step short-term power consumption forecasting with a hybrid deep learning strategy. *Energies* **2018**, *11*, 3089. [CrossRef]
28. Yan, K.; Li, W.; Ji, Z.; Qi, M.; Du, Y. A Hybrid LSTM Neural Network for Energy Consumption Forecasting of Individual Households. *IEEE Access* **2019**, *7*, 157633–157642. [CrossRef]
29. Peng, H.K.; Marculescu, R. Multi-scale compositionality: Identifying the compositional structures of Social dynamics using deep learning. *PLoS ONE* **2015**, *10*, 1–28. [CrossRef]
30. Chen, Z.; Li, C.; Sanchez, R. Gearbox Fault Identification and Classification with Convolutional Neural Networks. *Shock Vib.* **2015**, *2015*, 390134. [CrossRef]
31. Wang, K.; Qi, X.; Liu, H. A comparison of day-ahead photovoltaic power forecasting models based on deep learning neural network. *Appl. Energy* **2019**, *251*, 113315. [CrossRef]
32. Gupta, J.N.D.; Sexton, R.S. Comparing backpropagation with a genetic algorithm for neural network training. *Omega* **1999**, *27*, 679–684. [CrossRef]
33. Kingma, D.P.; Ba, J.L. Adam: A method for stochastic optimization. In Proceedings of the 3rd International Conference on Learning Representations, ICLR 2015—Conference Track Proceedings, San Diego, CA, USA, 7–9 May 2015; pp. 1–15.
34. Hu, Q.; Zhang, R.; Zhou, Y. Transfer learning for short-term wind speed prediction with deep neural networks. *Renew. Energy* **2016**, *85*, 83–95. [CrossRef]
35. Liu, S.; Ji, H.; Wang, M.C. Nonpooling Convolutional Neural Network Forecasting for Seasonal Time Series With Trends. *IEEE Trans. Neural Netw. Learn. Syst.* **2019**, 1–10. [CrossRef] [PubMed]

Article

Vector Speed Regulation of an Asynchronous Motor Based on Improved First-Order Linear Active Disturbance Rejection Technology

Xuesong Zhou *, Chenglong Wang * and Youjie Ma

Tianjin Key Laboratory for Control Theory and Application in Complicated Systems,
Tianjin University of Technology, Tianjin 300384, China; zhaohanyu@email.tjut.edu.cn
* Correspondence: eeliwei@email.tjut.edu.cn (X.Z.); 183125328@stud.tjut.edu.cn (C.W.);
  Tel.: +86-188-3026-7211(C.W.)

Received: 8 March 2020; Accepted: 25 April 2020; Published: 1 May 2020

Abstract: Asynchronous motors are widely used in industry and agriculture because of their simple structure, low cost, and easy maintenance. However, due to the coupling and uncertain factors of the actual operation of the motor, a traditional controller cannot achieve a satisfactory control effect. A linear active disturbance rejection controller (LADRC), featuring good robustness and adaptability, was proposed to improve the control efficiency of a nonlinear, uncertain plant. A linear extended state observer (LESO) is the core part of a L. The accuracy of the observation of state variables and unknown disturbances is related to the control effect of the controller. The performance of a traditional LESO is not high enough, and thus an error differential is introduced by analyzing the principle of LESO to improve its observation performance. The improved LADRC applies to the vector speed control of the induction motor. Additionally, low-speed and high-speed no-load starting, sudden load, electromagnetic torque, and three-phase stator current of the induction motor was simulated using MATLAB (Developed by MathWorks in Natick, MA, USA, and dealt by MathWorks Software (Beijing) Co., Ltd. in Beijing, China). Theoretical analysis and simulation results show that the ADRC based on the improved linear expansion observer was better than the traditional linear ADRC in terms of the dynamic and static performance and robustness.

Keywords: asynchronous motor; linear active disturbance rejection control; error differentiation; vector control

---

1. Introduction

Three-phase asynchronous motors are commonly used in industrial and agricultural production due to their advantages of simple structure, ruggedness, and low price [1–4]. The emergence of vector control allows the AC speed regulation system to have good speed regulation performance, just like a DC speed regulation system [5]. However, vector control has disadvantages, such as dependence on accurate mathematical models, poor adaptability to instruction changes, and sensitivity to changes in the system parameters. Even if the motor parameters and rotor flux linkage are known accurately, decoupling can be achieved under steady-state conditions, and there are still couplings during the field-weakening speed regulation. The nonlinearity of the magnetization curve of the ferromagnetic material in the motor [6] leads to the nonlinearity of the motor inductance, and the change of the inductance parameter reduces the speed control effect of the vector control. The AC control system has the characteristics of non-linearity [1,4], strong couplings, and having multiple variables, which means the traditional control method based on a precise mathematical model faces severe challenges.

The active disturbance rejection control (ADRC) theory was proposed for the control of nonlinear uncertain systems. The system is linearized by compensating for the observed total disturbance.

The compensated system can be converted into an integrator series independent of whether the object is deterministic, linear, or nonlinear, or whether it is time-varying or time-invariant. At present, the ADRC theory has been applied to a six-rotor aircraft [7], a permanent magnet synchronous motor [8], DC–DC boost converter [9], and other fields. Abdul-Adheem et al. [10] applied the improved ADRC to the decoupling control of multivariable systems. The coupling is divided into two parts: static coupling (control input of the system) and dynamic coupling (parts other than the control input to the system). All ADRC pass a reversible static matrix (approximately reversible also applies) that is used for decoupling. Compared with the traditional decentralized control method (an automatic disturbance rejection controller designed independently for each part of the system), the improved ADRC decoupling algorithm uses part of the system model information, easing the burden of the extended state observer (ESO) such that the decoupling effect is better. Although both numerical simulations and physical verification show that the ADRC controller has a good control effect, the large number of modeled non-linear links between the ADRC require high system hardware requirements and increases the difficulty of real-time control. The second-order auto-disturbance rejection controller has 15 parameters that need to be adjusted, and the direction of the parameter adjustment is difficult to determine, which brings certain difficulties in the practical application of the controller. In short, many factors limit the popularity and engineering application of ADRC.

American scholar Gao Zhiqiang explored the connotation and meaning of the idea of auto-disturbance control. He was inspired by the concept of "time scale" [11] proposed by Han Jingqing researchers and proposed the concept of "frequency scale." The parameter setting is carried out through the pole configuration in the frequency domain, and the parameters to be set are reduced to three, which greatly promotes the development and application of the auto-disturbance control theory. Since then, the linear auto-rejection controller has been used in fault detection [12], a wind energy conversion system [13], maximum power point tracking [14], and other fields. Li et al. [15] adopted the concept of "relative order" to determine the order of the linear ADRC (LADRC) controller and designed a second-order LADRC controller to suppress the harmonics to the grid in the AC microgrid. Laghridat et al. [13] applied LADRC to the control of generators and grid-side converters. Compared with the ADRC, an LADRC has the advantages of a fixed structure, an independent object model, clear physical meaning, easy theoretical analysis [13,16], and easy engineering application. However, in practical applications, it is found that the anti-interference performance of LADRC decreases rapidly with the increase of interference and input frequency, which is related to the insufficient performance of a traditional ESO [17].

This study took the three-phase cage asynchronous motor as the control object, established a mathematical model for it according to the rotor flux linkage orientation, and introduced the structure of the ADRC controller and the role of each component. First, the structure and function of each part of the ADRC controller are introduced. Then, based on the basic principle of deviation control, the adjustment process of each state variable of traditional LESO is discussed and improved. Through theoretical analysis, the stability proof and precision analysis of an improved LESO are given. From the frequency domain, the convergence, tracking, and immunity of the improved linear auto-disturbance controller are analyzed. In the time domain, a large deviation band of the initial value of the internal state variable of the observer is seen and the corresponding value of the system's overshooting is compared. Finally, the control effects of the two controllers are compared based on results from Matlab/Simulink digital simulation software (Developed by MathWorks in Natick, MA, USA, and dealt by MathWorks Software (Beijing) Co., Ltd. in Beijing, China).

2. Mathematical Modeling of an Asynchronous Motor and Introduction to Classic LADRC

2.1. Mathematical Modeling of the Rotor Flux Orientation of Induction Motor

To facilitate the research, it was necessary to treat the motor as an ideal motor; therefore, it was necessary to make the following assumptions [18]:

1. The stator and rotor three-phase windings of the motor are completely symmetrical.
2. The surfaces of the stator and rotor are smooth without any cogging effect, and the air gap magnetic potential of each phase of the stator and rotor exhibits a sinusoidal distribution in space.
3. The influences of the core eddy current, saturation, and hysteresis loss are ignored, and the skin effect of the conductor is ignored. (Note: the parameters of the rotor side have been converted to those of the stator side.)

The three-phase winding voltage balance equation in the three-phase stationary coordinate system is shown in Figure 1:

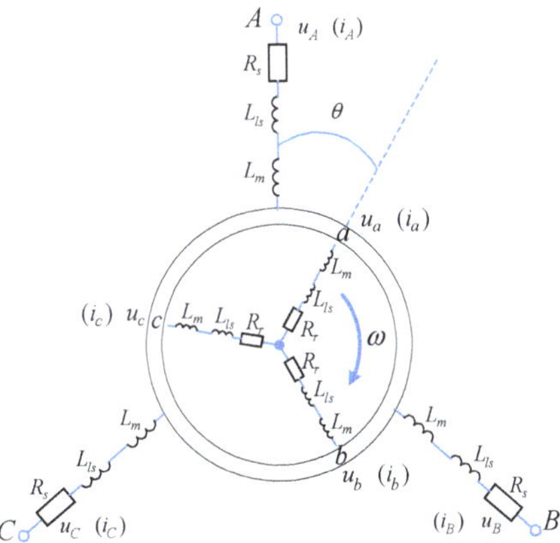

Figure 1. Physical model of an asynchronous motor.

The three-phase winding voltage balance equation in the three-phase stationary coordinate system is shown in Equation (1):

$$
\begin{bmatrix} u_A \\ u_B \\ u_C \\ u_a \\ u_b \\ u_c \end{bmatrix} = \begin{bmatrix} R_s & 0 & 0 & 0 & 0 & 0 \\ 0 & R_s & 0 & 0 & 0 & 0 \\ 0 & 0 & R_s & 0 & 0 & 0 \\ 0 & 0 & 0 & R_r & 0 & 0 \\ 0 & 0 & 0 & 0 & R_r & 0 \\ 0 & 0 & 0 & 0 & 0 & R_r \end{bmatrix} \begin{bmatrix} i_A \\ i_B \\ i_C \\ i_a \\ i_b \\ i_c \end{bmatrix} + \frac{d}{dt} \begin{bmatrix} \psi_A \\ \psi_B \\ \psi_C \\ \psi_a \\ \psi_b \\ \psi_c \end{bmatrix},
\tag{1}
$$

where u_A, u_B, u_C, u_a, u_b and u_c are the instantaneous values of the stator (A, B, C) and rotor phase (a, b, c) voltages; i_A, i_B, i_C, i_a, i_b, and i_c are the instantaneous values of the stator (A, B, C) and rotor (a, b, c) phase currents; and $\psi_A, \psi_B, \psi_C, \psi_a, \psi_b, \psi_c$ are the full flux of each phase winding. R_s and R_r are the resistances of the stator and rotor windings, respectively.

The flux of each winding of an asynchronous motor is the sum of its own self induction flux and the mutual inductance flux of other windings. Therefore, the flux of the six windings can be expressed as follows:

$$
\begin{bmatrix} \psi_s \\ \psi_r \end{bmatrix} = \begin{bmatrix} L_{ss} & L_{sr} \\ L_{rs} & L_{rr} \end{bmatrix} \begin{bmatrix} i_s \\ i_r \end{bmatrix},
\tag{2}
$$

where $\psi_s = \begin{bmatrix} \psi_A & \psi_B & \psi_C \end{bmatrix}^T, \psi_r = \begin{bmatrix} \psi_a & \psi_b & \psi_c \end{bmatrix}^T, i_s = \begin{bmatrix} i_A & i_B & i_C \end{bmatrix}^T$, and $i_r = \begin{bmatrix} i_a & i_b & i_c \end{bmatrix}^T$.

The stator inductance matrix is:

$$
\mathbf{L}_{ss} = \begin{bmatrix} L_{ms} + L_{ls} & -\frac{1}{2}L_{ms} & -\frac{1}{2}L_{ms} \\ -\frac{1}{2}L_{ms} & L_{ms} + L_{ls} & -\frac{1}{2}L_{ms} \\ -\frac{1}{2}L_{ms} & -\frac{1}{2}L_{ms} & L_{ms} + L_{ls} \end{bmatrix}.
\tag{3}
$$

The rotor inductance matrix is:

$$\mathbf{L}_{rr} = \begin{bmatrix} L_{ms} + L_{lr} & -\frac{1}{2}L_{ms} & -\frac{1}{2}L_{ms} \\ -\frac{1}{2}L_{ms} & L_{ms} + L_{lr} & -\frac{1}{2}L_{ms} \\ -\frac{1}{2}L_{ms} & -\frac{1}{2}L_{ms} & L_{ms} + L_{lr} \end{bmatrix}. \tag{4}$$

The mutual inductance matrix of the stator and rotor is:

$$\mathbf{L}_{rs} = \mathbf{L}_{sr}^{T} = L_{ms} \begin{bmatrix} \cos\theta & \cos(\theta - \frac{2\pi}{3}) & \cos(\theta + \frac{2\pi}{3}) \\ \cos(\theta + \frac{2\pi}{3}) & \cos\theta & \cos(\theta - \frac{2\pi}{3}) \\ \cos(\theta - \frac{2\pi}{3}) & \cos(\theta + \frac{2\pi}{3}) & \cos\theta \end{bmatrix}. \tag{5}$$

The voltage equation of the asynchronous motor in the rotating orthogonal coordinate system can be obtained from a Park transform:

$$\begin{bmatrix} u_{sd} \\ u_{sq} \\ u_{rd} \\ u_{rq} \end{bmatrix} = \begin{bmatrix} R_s & 0 & 0 & 0 \\ 0 & R_s & 0 & 0 \\ 0 & 0 & R_r & 0 \\ 0 & 0 & 0 & R_r \end{bmatrix} \begin{bmatrix} i_{sd} \\ i_{sq} \\ i_{rd} \\ i_{rq} \end{bmatrix} + \frac{d}{dt} \begin{bmatrix} \psi_{sd} \\ \psi_{sq} \\ \psi_{rd} \\ \psi_{rq} \end{bmatrix} + \begin{bmatrix} -\omega_1\psi_{sq} \\ \omega_1\psi_{sd} \\ -(\omega_1 - \omega)\psi_{rq} \\ (\omega_1 - \omega)\psi_{rd} \end{bmatrix}, \tag{6}$$

where u_{sd}, u_{sq}, u_{rd}, and u_{rq} are the components of the voltage of the stator and rotor sides along the d and q axes; i_{sd}, i_{sq}, i_{rd}, i_{rq} are the components of the current of the stator and rotor sides along the d and q axes; ψ_{sd}, ψ_{sq}, ψ_{rd}, and ψ_{rq} are the components of the magnetic flux of the stator and rotor sides along the d and q axes; and ω_1 and ω are the synchronous angular velocity and the rotor angular velocity, respectively.

The flux equation of the asynchronous motor in the synchronous rotation orthogonal coordinates system is:

$$\begin{bmatrix} \psi_{sd} \\ \psi_{sq} \\ \psi_{rd} \\ \psi_{rq} \end{bmatrix} = \begin{bmatrix} L_s & 0 & L_m & 0 \\ 0 & L_s & 0 & L_m \\ L_m & 0 & L_r & 0 \\ 0 & L_m & 0 & L_r \end{bmatrix} \begin{bmatrix} i_{sd} \\ i_{sq} \\ i_{rd} \\ i_{rq} \end{bmatrix}, \tag{7}$$

where L_s, L_r, and L_m are stator inductance, rotor inductance, and mutual inductance between the stator and rotor, respectively. $\psi_{rd} = \psi_r$ when the rotor flux is oriented as in Figure 2.

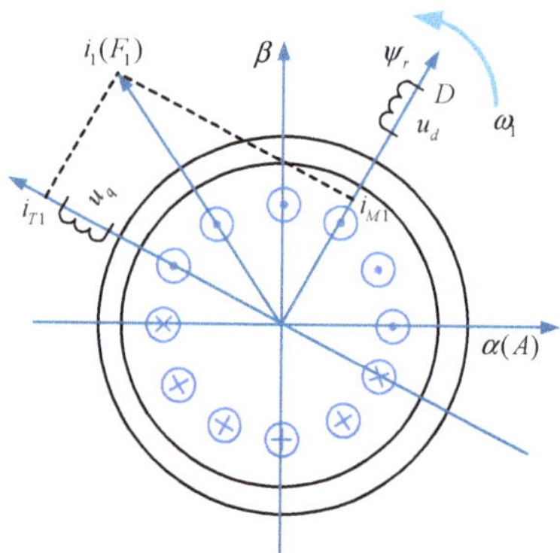

Figure 2. Rotor flux orientation.

Selecting the stator currents and rotor flux as the state variables, and the stator voltages as the control variables, a fourth-order simplified nonlinear differential equation can be given as follows:

$$\begin{cases} \dot{\omega}_r = k_1 \psi_r i_{sq} - \dfrac{n_p}{J} T_L, \\ \dot{\psi}_{rd} = \dfrac{L_m}{T_r} i_{sd} - \dfrac{1}{T_r} \psi_r, \\ \dot{i}_{sd} = -k_2 i_{sd} + k_3 \psi_r + \omega_1 i_{sq} + \dfrac{1}{\sigma L_s} u_{sd}, \\ \dot{i}_{sq} = -k_2 i_{sq} - \dfrac{L_m}{\sigma L_s L_r} \psi_r \omega_r - \omega_1 i_{sd} + \dfrac{1}{\sigma L_s} u_{sq}. \end{cases} \tag{8}$$

In the above state equations, n_p, T_r, and T_L represent the number of pole pairs of the motor, the load torque, and the rotor time constant, respectively. The coupling between the state variables in Equation (8) causes the nonlinearity of the system [4,10]. The vector control block diagram of the asynchronous motor is shown in Figure 3, in which automatic speed regulator (*ASR*), the automatic current torque regulator (*ACTR*), and the automatic current magnetic regulator (*ACMR*) are the controllers of the vector speed regulation system.

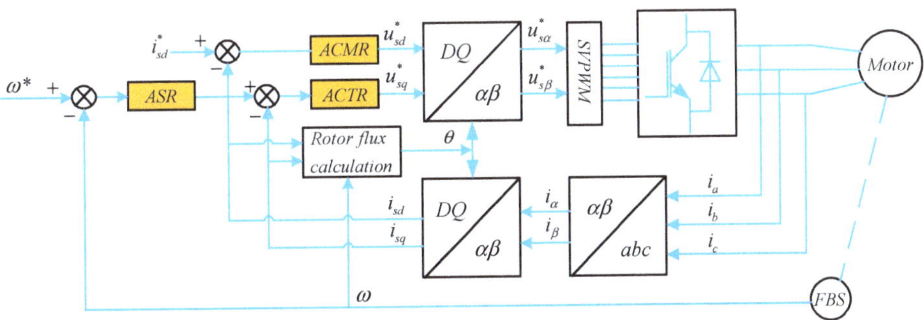

Figure 3. Vector control structure of an asynchronous motor. ASR: automatic speed regulator, ACTR: Automatic torque current regulator, ACMR: Automatic magnetic field current regulator SVPWM: Space Vector Pulse Width Modulation.

In Equation (8), there is a cross-coupling term in the state equation of the asynchronous motor, thus resulting in the mutual effect of control of the torque component and the excitation component of the stator current, which further affects the dynamic and static performance of the system. The graphical representation of the coupling term is shown in Figure 4. Decoupling can be achieved if the coupling terms are observed and compensated for [19].

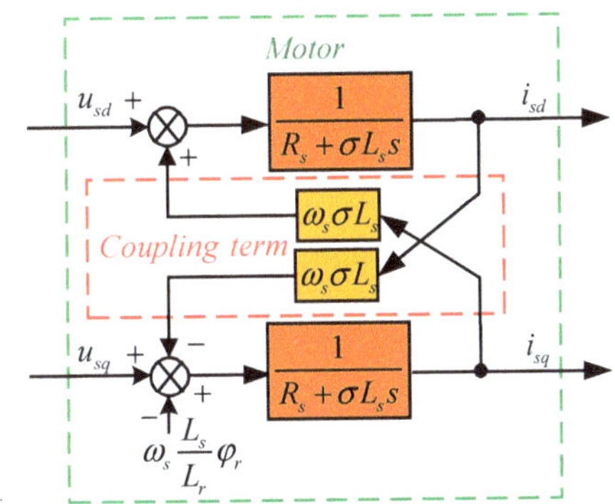

Figure 4. Coupling structure of an asynchronous motor.

2.2. LADRC Introduction and First-Order LADRC Design

The ADRC consists of a tracking differentiator (TD) [20,21], an extended state observer (ESO) [22,23], and a nonlinear states error feedback control laws (NLSEF) . Among them, the TD can arrange the transition process for the given input, reduce the "impact" of the original error on the system caused by the given mutation, and realize the differential signal extraction; ESO can observe the state variables and total disturbances (unmodeled dynamics and external disturbances) of the controlled object in a real-time manner; and the NLSEF is used to improve the dynamic characteristics of the closed-loop system. The linear ADRC only replaces the nonlinear part of the original controller with the linear part, where the structure of the controller is shown in Figure 5.

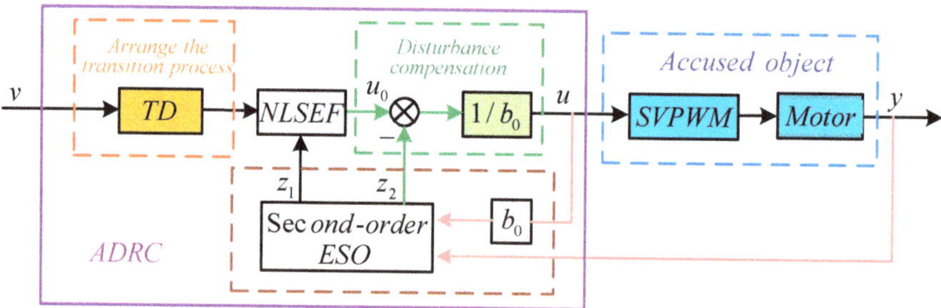

Figure 5. Structure of an active disturbance rejection controller (TD: tracking differentiator, NLSEF: nonlinear state error feedback, ESO: extended state observer).

Consider the first-order system [24]:

$$\dot{y} = -ay + w + bu, \tag{9}$$

where u is the output of the controller, y is the system output, w is the external disturbance, a is the system parameter, b is the controller gain that satisfies $b \approx b_0$; the parameters a and b are unknown. Let $x_1 = y$ and $x_2 = f(y, w) = -ay + w + (b - b_0)u$, where x_1 represents the system output and x_2 represents the total disturbance of the system. Assuming that $f(y, w)$ is derivable and satisfies $\dot{f}(y, w) = h$, the state variables $x_1 = y$ and x_2 are selected to establish the continuous extended state equation s shown in Equation (10):

$$\begin{cases} \begin{bmatrix} \dot{x}_1 \\ \dot{x}_2 \end{bmatrix} = \begin{bmatrix} 0 & 1 \\ 0 & 0 \end{bmatrix} \begin{bmatrix} x_1 \\ x_2 \end{bmatrix} + \begin{bmatrix} b_0 \\ 0 \end{bmatrix} u + \begin{bmatrix} 0 \\ 1 \end{bmatrix} \dot{f}, \\ y = \begin{bmatrix} 1 & 0 \end{bmatrix} \begin{bmatrix} x_1 \\ x_2 \end{bmatrix}. \end{cases} \tag{10}$$

The corresponding continuous expansion state observer can be established as:

$$\begin{cases} e = z_1 - y, \\ \begin{bmatrix} \dot{z}_1 \\ \dot{z}_2 \end{bmatrix} = \begin{bmatrix} -\beta_1 & 1 \\ -\beta_2 & 0 \end{bmatrix} \begin{bmatrix} z_1 \\ z_2 \end{bmatrix} + \begin{bmatrix} b_0 & \beta_1 \\ 0 & \beta_2 \end{bmatrix} \begin{bmatrix} u \\ y \end{bmatrix}. \end{cases} \tag{11}$$

In Equation (11), z_1 and z_2 are the state variables of the linear extended state observer, which can be adjusted using the difference between the state variables z_1 and the system output y. By selecting the appropriate observer gain coefficients β_1 and β_2, the observer state variables can be used to observe the system output y and the total disturbance $f(y, w)$.

Estimate the total disturbance of the system through the expanded state variables, and compensate for the input side of the system:

$$u = \frac{(-z_2 + u_0)}{b_0}.$$ (12)

If the estimation error of z_2 to $f(y, w)$ is excluded, Equation (9) is simplified to a pure integral link:

$$\dot{y} = (f(y, w) - z_2) + u_0 \approx u_0.$$ (13)

The linear feedback control law uses proportional links:

$$u_0 = kp(v - z_1).$$ (14)

In Equation (14), kp is the controller bandwidth and v is the given reference input. Pole assignment of the parameters of the observer and controller is performed using the bandwidth method [10]:

$$\begin{cases} \beta_1 = 2\omega_0, \\ \beta_2 = \omega_0^2, \\ k_p = \omega_c. \end{cases}$$ (15)

where ω_c is the controller gain and ω_0 is the observer gain. The bandwidth ω_0 of the observer is about 3 to 5 times that of the controller ω_c.

3. Design and Performance Analysis of ADRC

3.1. Design and Stability Proof of the Improved State Observer

The traditional observer estimates the internal state variables of the system using the difference e between the estimated value z_1 of the system output and the output y of the system. In Sun [25], it is pointed out that the observer should first track the output y of the system with z_1, and then track the output f with z_2. Before the tracking of the estimated value z_1 to the system output y is completed, other state variables of the observer cannot complete the tracking of the corresponding state variables of the system. However, when the observation variable z_1 can better track the output y of the system, the smaller error e makes it difficult to adjust other observation variables; therefore, we have to use a larger coefficient β_2 to speed up the tracking of the observation variables to the real value. Meanwhile, to ensure the stability of the system, the value of β_2 cannot be too large. In general, the gain coefficient of the observer β_i increases by an order of magnitude, which is more serious in the higher-order ADRC. Yang et al. [26] mentioned that the extended observer could be improved by introducing the differential of the observation error, but the parameters of the controller were doubled and the parameters of the observer needed to be configured; therefore, it had to be further simplified.

From Equation (11), we can get:

$$\begin{cases} z_1 = x_1 + e, \\ z_2 = \dot{z}_1 + \beta_1 e - b_0 u. \end{cases}$$ (16)

The following equation can be obtained by sorting Equation (16):

$$\begin{cases} z_1 = x_1 + e, \\ z_2 = x_2 + \dot{e} + \beta_1 e. \end{cases}$$ (17)

It can be seen from Equation (17) that the error between z_2 and x_2 is $\dot{e} + \beta_1 e$, and adjusting z_2 by using it as a correction amount can speed up the convergence without significantly increasing the observer gain. Therefore, the classic LESO can be modified as follows:

$$\begin{cases} e = z_1 - x_1, \\ \dot{z}_1 = z_2 - \beta_1 e + b_0 u, \\ \dot{z}_2 = -\beta_2(\dot{e} + \beta_1 e). \end{cases} \tag{18}$$

Equations (12), (14), and (18) form the improved Linear Active Disturbance Rejection Controller of Equation (8), whose structure is shown in Figure 6.

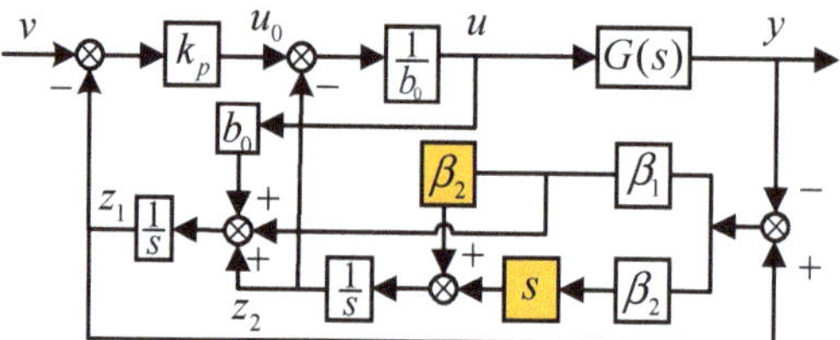

Figure 6. Structure of the improved controller.

3.1.1. Improved Stability Proof of LESO

Let $e_1 = z_1 - x_1$ and $e_2 = z_2 - x_2$. From Equations (10) and (18), we obtain:

$$\begin{cases} \dot{e}_1 = e_2 - \beta_1 e_1, \\ \dot{e}_2 = -\beta_2 e_2 - w. \end{cases} \tag{19}$$

Let $Y_1 = e_1$ and $Y_2 = e_2 - \beta_1 e_1$ to obtain the error equation of the LESO system:

$$\begin{cases} \dot{Y}_1 = Y_2, \\ \dot{Y}_2 = -(\beta_1 + \beta_2)Y_2 - \beta_1 \beta_2 Y_1 - w. \end{cases} \tag{20}$$

The characteristic equation of Equation (20) is:

$$\lambda^2 + (\beta_1 + \beta_2)\lambda + \beta_1 \beta_2 = 0. \tag{21}$$

The necessary and sufficient conditions for the stability of the second-order system are $\beta_1 + \beta_2 > 0$ and $\beta_1 \beta_2 > 0$. The zero solution ($e_1 = 0$, $e_2 = 0$) of the second-order constant-coefficient differential equation shown in Equation (20) is globally asymptotically stable because $\omega_0 > 0$ and $\omega_c > 0$ are stable.

When considering the disturbance w, the system has a steady-state error. Specify $w_0 = const > 0$ when $|w| \leq w_0$. When the system reaches a steady-state, then:

$$\begin{cases} \dot{Y}_1 = Y_2 = 0, \\ \dot{Y}_2 = 0. \end{cases} \tag{22}$$

The steady-state error is calculated according to Equation (19):

$$\begin{cases} |e_1| \leq \frac{w_0}{\beta_1 \beta_2}, \\ |e_2| \leq \frac{w_0}{\beta_2}. \end{cases} \tag{23}$$

3.1.2. Observation Errors of the Classical First-Order LESO

Stability and error analyses are performed on the traditional first-order LESO represented by Equation (11). Let $Y_1 = e_1$ and $Y_2 = e_2 - \beta_1 e_1$ be used to obtain the equation of the traditional LESO error system:

$$\begin{cases} \dot{Y}_1 = Y_2, \\ \dot{Y}_2 = -\beta_1 Y_2 - \beta_2 Y_1 - w. \end{cases} \tag{24}$$

The characteristic equation of Equation (24) is given as follows:

$$\lambda^2 + \beta_1 \lambda + \beta_2 = 0. \tag{25}$$

Using the Hurwitz theorem, the necessary and sufficient conditions for the stability of the second-order system are $\beta_1 > 0$ and $\beta_2 > 0$. The zero solution ($e_1 = 0$, $e_2 = 0$) of the second-order constant-coefficient differential equation shown in Equation (24) is globally asymptotically stable because $\omega_0 > 0$ and $\omega_c > 0$ are stable.

When considering the disturbance w, the system has a steady-state error. Specify $w_0 = const > 0$ when $|w| \leq w_0$. When the system reaches a steady-state, then:

$$\begin{cases} \dot{Y}_1 = Y_2 = 0, \\ \dot{Y}_2 = 0. \end{cases} \tag{26}$$

The steady-state error of the observer can be expressed as:

$$\begin{cases} |e_1| \leq \frac{w_0}{\beta_2}, \\ |e_2| \leq \frac{\beta_1 w_0}{\beta_2}. \end{cases} \tag{27}$$

According to the above analysis, the modified LESO shown in Equation (18) can exhibit a better dynamic regulation performance and a smaller steady-state observation error than the traditional LESO when the parameters β_1 and β_2 are the same. Compared with Equations (23) and (27), the improved LESO exhibits a higher observation accuracy than the traditional LESO when the observer and controller bandwidths are the same.

3.2. Performance Index Analysis of the Improved Linear ADRC

3.2.1. Convergence and Estimation Error of the Improved LESO

The Laplace transform of Equation (18) can be used to obtain the transfer function of the observer:

$$\begin{cases} Z_1(s) = \frac{(\beta_1+\beta_2)s+\beta_1\beta_2}{(s+\beta_1)(s+\beta_2)} Y(s) + \frac{b_0 s}{(s+\beta_1)(s+\beta_2)} U(s), \\ Z_2(s) = \frac{\beta_2 s}{s+\beta_2} Y(s) - \frac{b_0 \beta_2}{s+\beta_2} U(s). \end{cases} \tag{28}$$

Taking $e_1 = z_1 - y$ and $e_2 = z_2 - \dot{y}$ into account for analyzing a typical y, and u as the amplitude K step signal, then the steady-state error of LESO is given as:

$$\begin{cases} e_1 = \lim_{s \to 0} sE_1(s) = 0, \\ e_2 = \lim_{s \to 0} sE_2(s) = 0. \end{cases} \tag{29}$$

Equation (29) shows that the improved LESO has a good convergence performance for realizing the invariant estimation of the system state variables and generalized disturbances. Further analyzing its dynamic process, the response of z_1 under the step signal when $b_0 = 0$ is as follows:

$$z_1(s) = K(\frac{1}{s} + \frac{2}{(s + 2\omega_0)(\omega_0 - 2)} - \frac{\omega_0}{(s + \omega_0^2)(\omega_0 - 2)}). \tag{30}$$

The time-domain response of z_1 under the action of a step signal can be obtained using an inverse Laplace transform:

$$z_1(t) = \begin{cases} K(1 + \frac{2e^{-2t\omega_0} - \omega_0 e^{-t\omega_0^2}}{\omega_0 - 2}) & \omega_0 \neq 2, \\ K(1 + 4te^{-4t} - e^{-4t}) & \omega_0 = 2. \end{cases} \tag{31}$$

In Equation (30), for $t > 0$, the derivative of t and taking $\dot{z}_1(t) = 0$ can produce the extreme point of t_0:

$$t_0 = \begin{cases} \frac{2(\ln\omega_0 - \ln 2)}{\omega_0^2 - 2\omega_0} & \omega_0 \neq 2, \\ \frac{1}{2} & \omega_0 = 2. \end{cases} \tag{32}$$

The extreme value of z_1 is obtained by substituting the extreme value t_0 into Equation (30):

$$z_1(t_0) = \begin{cases} K(1 + \frac{2e^{\frac{-4(\ln^{\omega}0 - \ln 2)}{\omega_0 - 2}} - \omega_0 e^{\frac{-2\omega_0(\ln^{\omega}0 - \ln 2)}{\omega_0 - 2}}}{\omega_0 - 2}) & \omega_0 \neq 2, \\ K(1 + e^{-2}) \approx 1.135K & \omega_0 = 2. \end{cases} \tag{33}$$

The trajectory of $z_1(t_0)$ with the observer bandwidth value ω_0 can be obtained via digital simulation. According to Figure 7, when $\omega_0 = 2$, the tracking overshoot of observer $z_1(t_0)$ to y is the largest. At this time, the system overshoot is equal to the traditional LESO overshoot. The traditional LESO has a 13.5% overshoot at $t_0 = 2/\omega_0$ and the amount of overshoot is independent of the value of the observer bandwidth ω_0.

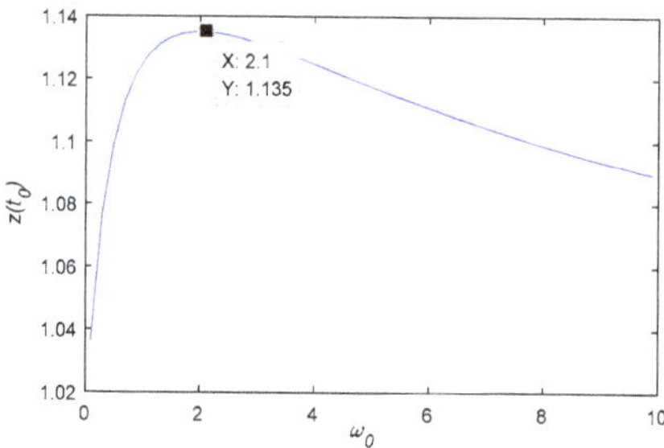

Figure 7. Relation curve between the system overshoot and observer bandwidth.

The overshoot of the improved LESO varies with the observer bandwidth ω_0. The maximum overshoot of the observer state variable z_1 is equal to that of the traditional observer. The corresponding speed of the system can be increased and the amount of overshoot is reduced by selecting a larger observer bandwidth ω_0. Although the observer bandwidth ω_0 is larger and the tracking speed is faster, it will lead to noise amplification. The ability of LESO to suppress noise needs to be analyzed.

3.2.2. Improved Disturbance Immunity Analysis of LESO

The closed-loop transfer function of the improved LADRC can be obtained by combining Equations (12), (14), and (18):

$$u = \frac{1}{b_0} G_1(s)(k_p v - H(s)y). \tag{34}$$

The transfer functions of $G_1(s)$ and $H(s)$ in Equation (34) are as follows:

$$
\begin{aligned}
G_1(s) &= \frac{(s+\beta_1)(s+\beta_2)}{s^2+(\beta_1+k_p)s}, \\
H(s) &= \frac{\beta_2 s^2+(\beta_1\beta_2+\beta_1 k_p+\beta_2 k_p)s+\beta_1\beta_2 k_p}{(s+\beta_1)(s+\beta_2)}.
\end{aligned}
\tag{35}
$$

According to Equation (34), the diagram of the system structure shown in Figure 8 is obtained [24]:

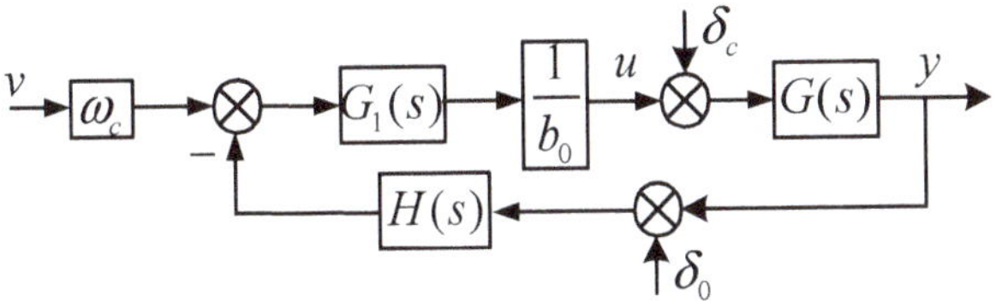

Figure 8. Equivalent system structure diagram.

Now the effect of the observation noise δ_0 at the output y of the system and the disturbance δ_c at the output u of the controller for the improved LESO will be discussed. Based on Equation (28), the transfer function of the improved LADRC is:

$$
\frac{z_1}{\delta_0} = \frac{(2\omega_0 + \omega_0^2)s + 2\omega_0^3}{(s + 2\omega_0)(s + \omega_0^2)}.
\tag{36}
$$

Similarly, the transfer function of δ_0 of the traditional LESO's observation noise can be obtained as follows:

$$
\frac{z_1}{\delta_0} = \frac{2\omega_0 s + \omega_0^2}{(s + \omega_0)^2}.
\tag{37}
$$

The transfer function of the disturbance δ_c at the output of the LESO controller can be improved according to:

$$
\frac{z_1}{\delta_c} = \frac{b_0 s}{(s + 2\omega_0)(s + \omega_0^2)}.
\tag{38}
$$

The transfer function of the disturbance δ_c at the output of a traditional LESO controller is:

$$
\frac{z_1}{\delta_c} = \frac{b_0 s}{(s + w_0)^2}.
\tag{39}
$$

Figure 9 shows the characteristic amplitude and phase–frequency curves for the improved and traditional LESOs. The bandwidth of the improved LESO was higher than that of the traditional LESO, and the phase lag of the intermediate frequency segment was improved. Unlike Figure 9, the improved LESO with the same observer bandwidth in Figure 10 is basically the same as the traditional LESO in the high frequency band, but it has better noise immunity in the low frequency band compared with the traditional LESO, and can more effectively suppress the interference at the input end.

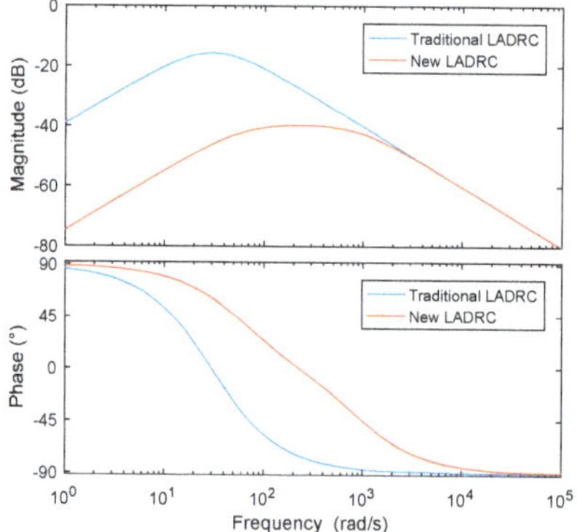

Figure 9. Frequency-doman characteristic curves of the observed noise.

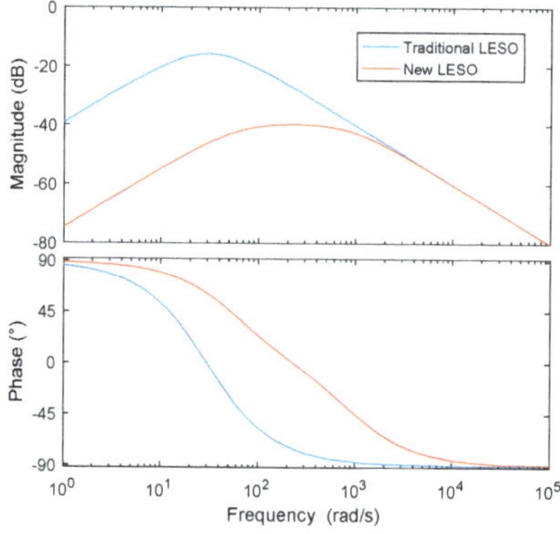

Figure 10. Frequency-domain characteristic curves of the input disturbance.

3.2.3. Immunity Analysis of the Improved Self-Disturbance Rejection Controller

According to Equation (10), the control object can be written as:

$$sY(s) = F(s) + b_0U(s). \tag{40}$$

Combined with Figure 8, the closed-loop transfer function of the system is as follows:

$$Y(s) = \frac{k_p}{s + k_p}V(s) + \frac{s^2 + (\beta_1 + k_p)s}{(s + \beta_1)(s + \beta_2)(s + k_p)}F(s). \tag{41}$$

The closed-loop transfer function of the system includes the tracking term and the disturbance term. If the state variable of the observer can be used to accurately estimate the total disturbance of the system, the closed-loop transfer function of the system is simplified to the first-order inertial link. At this time, it relates the corresponding speed of the system to the bandwidth of the controller, and the larger the bandwidth, the faster the system.

It can be seen from the closed-loop transfer function that the disturbance term impacts the observer and controller bandwidths. The same observer and controller bandwidths were selected for comparing the improved LESO with the traditional LESO. It can be seen from Figure 11 that under the same bandwidth, the immunity of the improved LESO in the middle- and low-frequency bands were better than that of the traditional LESO, and it improved the phase lag of the middle-frequency band.

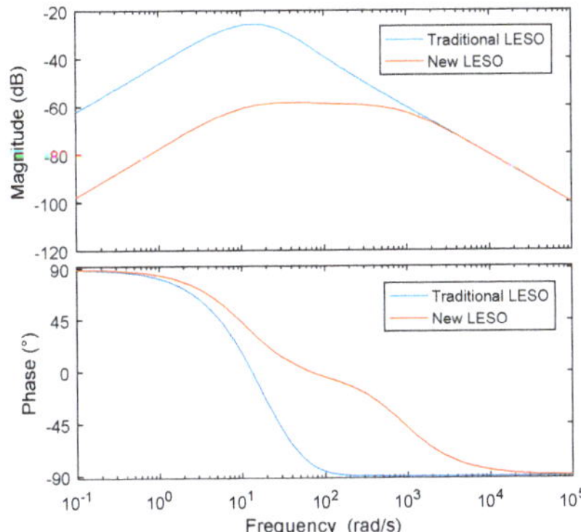

Figure 11. Logarithmic phase–frequency characteristic curves of the disturbance term.

In particular, if the disturbance f is taken as the unit step signal, we can obtain the output response using Equation (41):

$$Y(s) = \frac{a}{(s+\omega_0)} + \frac{b}{(s+\omega_0^2)} + \frac{c}{(s+\omega_c)},\tag{42}$$

$$\begin{cases} a = \dfrac{\omega_0+\omega_c}{(\omega_0^2-\omega_0)(\omega_c-\omega_0)}, \\ b = \dfrac{2\omega_0+\omega_c-\omega_0^2}{(\omega_0-\omega_0^2)(\omega_c-\omega_0^2)}, \\ c = \dfrac{2\omega_0}{(\omega_0-\omega_c)(\omega_0^2-\omega_c)}. \end{cases}\tag{43}$$

The time domain for the anti-Laplace transformation to obtain system output can be expressed as:

$$y(t) = ae^{-\omega_0 t} + be^{-\omega_0^2 t} + ce^{-\omega_c t}.\tag{44}$$

It is easy to find $\lim\limits_{t\to\infty} y(t) = 0$, i.e., the steady-state output of the system is zero under the external step disturbance.

(a) Controller Stability with an Unknown Input Gain

Considering only the influence of the input gain on the stability of the system, i.e., if $f = (b - b_0)u$, then Equation (40) can be simplified to:

$$sY(s) = bU(s).\tag{45}$$

The following equation can be obtained by combining Equations (26) and (30):

$$Y(s) = \frac{\omega_c(s+2\omega_0)(s+\omega_0^2)}{a_3 s^3 + a_2 s^2 + a_1 s + a_0} V(s).\tag{46}$$

In Equation (46), the coefficients are as follows: $a_3 = b_0/b, a_2 = 2a_0\omega_0 + a_0\omega_c + \omega_0^2$, $a_1 = 2\omega_0^3 + 2\omega_0\omega_c + \omega_0^2\omega_c$, and $a_0 = 2\omega_0^3\omega_c$. As ω_0 and ω_c are greater than zero, it is easy to see that a_3, a_2, a_1 and a_0 are all positive numbers. The necessary and sufficient condition for the stability of the Leonard qipat stability criterion (Equation (46)) is that all odd or even Hurwitz determinants are positive.

$$\Delta_3 = \begin{vmatrix} a_2 & a_0 & 0 \\ a_3 & a_1 & 0 \\ 0 & a_2 & a_0 \end{vmatrix} = a_0(a_1 a_2 - a_0 a_3)$$
$$= a_0^2(4\omega_0^4 + 2\omega_0^3\omega_c + \omega_c^2\omega_c^2 + 4\omega_0^2\omega_c + 2\omega_0\omega_c^2) + (2\omega_0^5 + \omega_0^4\omega_c + 2\omega_0^3\omega_c)a_0 \tag{47}$$

Since b, b_0, ω_0 and ω_c are all positive numbers, $\Delta_3 > 0$ is true, i.e., the improved LADRC can be stable for any parameter greater than zero.

(b) Stability of the Controller when the System Parameters are Unknown

Set the controlled object as the following:

$$y = \frac{b_0}{s + k_c}u,\tag{48}$$

where k_c in Equation (48) is an unknown system parameter, and the closed-loop transfer function of the system is obtained by combining with Figure 8:

$$Y(s) = \frac{\omega_c(s + \omega_0)(s + \omega_0^2)(s + 2\omega_0)}{s^4 + a_4 s^3 + a_3 s^2 + a_2 s + a_1}V(s),\tag{49}$$

$$\begin{cases} a_4 = \omega_0^2 + 4\omega_0 + \omega_c + k_c, \\ a_3 = \omega_0^2\omega_c + 4\omega_0^2 + 3\omega_0^3 + 4k_c\omega_0 + k_c\omega_c + 4\omega_0\omega_c, \\ a_2 = 4k_c\omega_0^2 + 2\omega_0^2\omega_c + 3\omega_0^3\omega_c + 2\omega_0^4 + 2k_c\omega_0\omega_c, \\ a_1 = 2\omega_0^4\omega_c. \end{cases}\tag{50}$$

The necessary and sufficient conditions for the stability of Equation (49) are:

$$ak_c^3 + bk_c^2 + ck_c + d > 0,\tag{51}$$

$$\begin{cases} a = & 4\omega_0^5\omega_c(4\omega_0 + \omega_c)(2\omega_0 + \omega_c), \\ b = & (72\omega_0^9\omega_c + 68\omega_0^8\omega_c^2 + 128\omega_0^8\omega_c + 14\omega_0^7\omega_c^3 + 160\omega_0^7\omega_c^2 + 52\omega_0^6\omega_c^3 + 4\omega_0^5\omega_c^4, \\ c = & 52\omega_0^{11}\omega_c + 62\omega_0^{10}\omega_c^2 + 176\omega_0^{10}\omega_c + 16\omega_0^9\omega_c^3 + 252\omega_0^9\omega_c^2 + 128\omega_0^9\omega_c + \\ & 104\omega_0^8\omega_c^3 + 272\omega_0^8\omega_c^2 + 10\omega_0^7\omega_c^4 + 144\omega_0^7\omega_c^3 + 20\omega_0^6\omega_c^4, \\ d = & 12\omega_0^{13}\omega_c + 18\omega_0^{12}\omega_c^2 + 56\omega_0^{12}\omega_c + 6\omega_0^{11}\omega_c^3 + 96\omega_0^{11}\omega_c^2 + 64\omega_0^{11}\omega_c + \\ & 48\omega_0^{10}\omega_c^3 + 160\omega_0^{10}\omega_c^2 + 6\omega_0^9\omega_c^4 + 124\omega_0^9\omega_c^3 + 64\omega_0^9\omega_c^2 + 24\omega_0^8\omega_c^4 + \\ & 72\omega_0^8\omega_c^3 + 16\omega_0^7\omega_c^4. \end{cases}\tag{52}$$

If the roots of Equation (51) are k_{c1}, k_{c2}, and k_{c3} ($k_{c1} < k_{c2} < k_{c3}$), the conditions of system stability are $k_{c1} < k_c < k_{c2}$ or $k_c > k_{c3}$.

According to the digital simulation results, the equation had a pair of conjugate complex roots and a real root. Figure 12 shows the boundary curve for ensuring the system stability when $\omega_0 = 30$ and $\omega_0 \in [0, 60]$, and Figure 13 shows the boundary curve for ensuring the system stability when $\omega_0 = 30$ and $\omega_c \in [0, 60]$. From Figures 12 and 13, it can be seen that the stability region of the system increased with an increase of the observer bandwidth ω_0 and the controller bandwidth ω_c

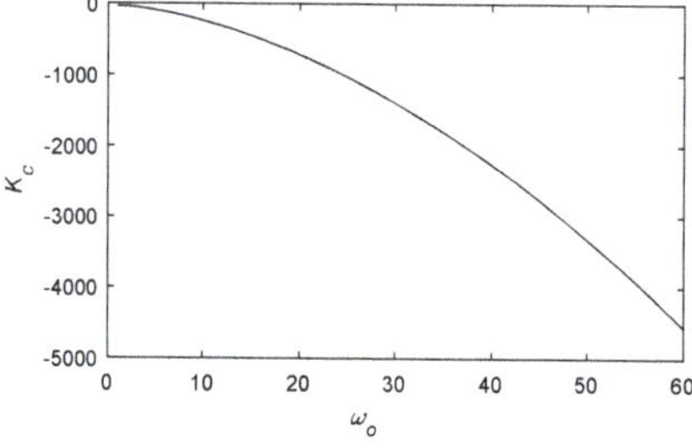

Figure 12. System stability region under different observer bandwidths ω_0.

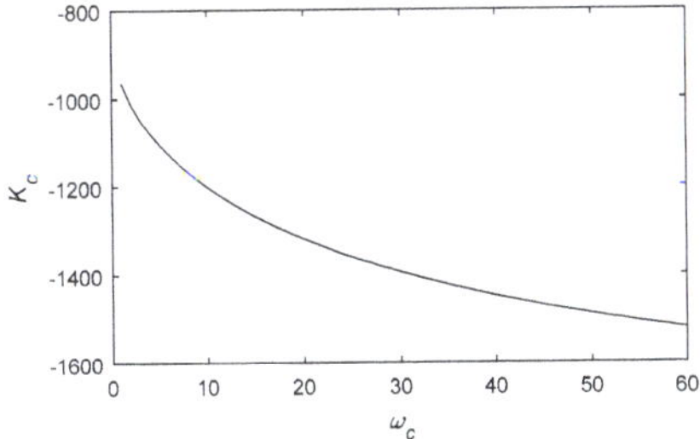

Figure 13. System stability region under different controller bandwidth ω_c.

4. Simulation Study

To verify the actual control effect of the improved LADRC , a vector control model of a three-phase asynchronous motor based on rotor flux linkage orientation was established based on MATLAB and Simulink simulation software (Development by MathWorks, Natick, MA, USA, and agent of MathWorks Software (Beijing) Co., Ltd.). The motor parameters are shown in Table A1. The current loop controller adopted the improved LADRC control, and the outer loop adopted Proportional integral controller (PI) control. The control effect of the controller was verified by simulating the motor speed, the three-phase stator current, the electromagnetic torque, and the sudden load when the motor was started without load at different speeds.

4.1. Dynamic Performance of the Controller for an Induction Motor at Different Given Speeds

4.1.1. Dynamic Performance of the Controller Given a Low Speed of the Motor

Figure 14a shows the simulation diagram of the no-load starting process of the asynchronous motor with a given reference speed of 200 rev/min. The red curve represents the improved LADRC, while the blue curve represents the traditional LADRC. It can be seen from Figure 14 that the traditional LADRC could reach the given value near the motor speed at 0.1 s, while the improved LADRC could reach the given value near the motor speed at 0.07 s. Therefore, the controller effect of the improved LADRC was better than that of the traditional LADRC.

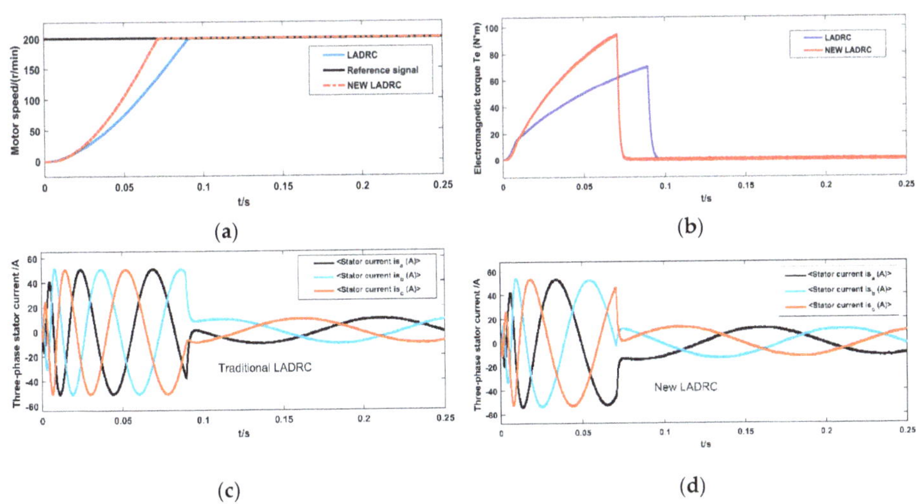

Figure 14. Low-speed simulation of an asynchronous motor. LADRC: Linear ADRC.

Figure 14b is a simulation diagram of the electromagnetic torque of the induction motor during no-load starting. The corresponding speed of the electromagnetic torque of the motor was greater when the improved LADRC was adopted. The electromagnetic torque of the motor when the improved LADRC was adopted was larger than that when the traditional LADRC was adopted at the same time point; therefore, the electromagnetic torque of the motor could be restored to zero in a shorter time.

Figure 14c is the simulation diagram of the three-phase stator current when the asynchronous motor started up without a load under the action of the traditional LADRC. After the improved LADRC was adopted, the three-phase stator current reached the steady-state again near 0.09 s, and the motor speed reached the given value. Figure 14d is the simulation diagram of the three-phase stator when the asynchronous motor started up without a load under the action of the improved LADRC. After the improved LADRC was adopted, the three-phase stator current reached the steady-state again near 0.07 s. It can be seen from the simulation diagram of the three-phase stator under the action of two controllers that the dynamic performance of the improved linear observer was better.

The improved linear expansion observer can better estimate the total disturbance in the system and realize the decoupling between the excitation subsystem and the torque subsystem. Therefore, the motor controlled by the improved Linear Active Disturbance Rejection Controller at the same time in Figure 14b can obtain a greater electromagnetic torque. The key to motor speed regulation is the adjustment of electromagnetic torque. At the same time in Figure 14b, the electromagnetic torque obtained by the motor under the improved linear auto-disturbance control is larger than that under the traditional linear auto-disturbance controller, so the improvement in Figure 14a Under the Linear Active Disturbance Rejection Controller, the motor speed can reach the given value of a shorter time.

The load in the equivalent circuit of the motor is purely resistive and the magnitude of the equivalent resistance is related to the slip rate. The slip rate is 1 when the motor is started without load. At this time, the total impedance of the system is small and the current on the stator side is large. When the motor reaches a given speed, the slip ratio is less than 1 and the total impedance of the system becomes larger. Therefore, the three-phase stator currents in Figure 14c,d become smaller when the motor speed reaches a given value.

4.1.2. Dynamic Performance of the Controller Given a High Speed of the Motor

Figure 15a is the speed simulation diagram of the asynchronous motor at a given reference speed of 800 rev/min, in which the blue curve is the simulation diagram under the action of the traditional LADRC, and the red curve is that under the action of the improved LADRC. It can be seen from the figure that the traditional LADRC reached the given value near the motor speed at 0.23 s, while the improved LADRC reaches the given value near the motor speed at 0.18 s, and exhibited a better control effect than the traditional LADRC.

Figure 15c is the simulation diagram of the three-phase stator when the asynchronous motor started up without a load under the action of the traditional LADRC. After the improved LADRC was adopted, the three-phase stator current reached the steady-state again near 0.17 s, when the motor speed reached the given value. Figure 15d is the simulation diagram of the three-phase stator when the asynchronous motor started up without a load under the action of the improved LADRC. After the improved LADRC was adopted, the three-phase stator current reached the steady-state again near 0.16 s. According to Figure 15c,d, the dynamic performance of the improved linear observer was better.

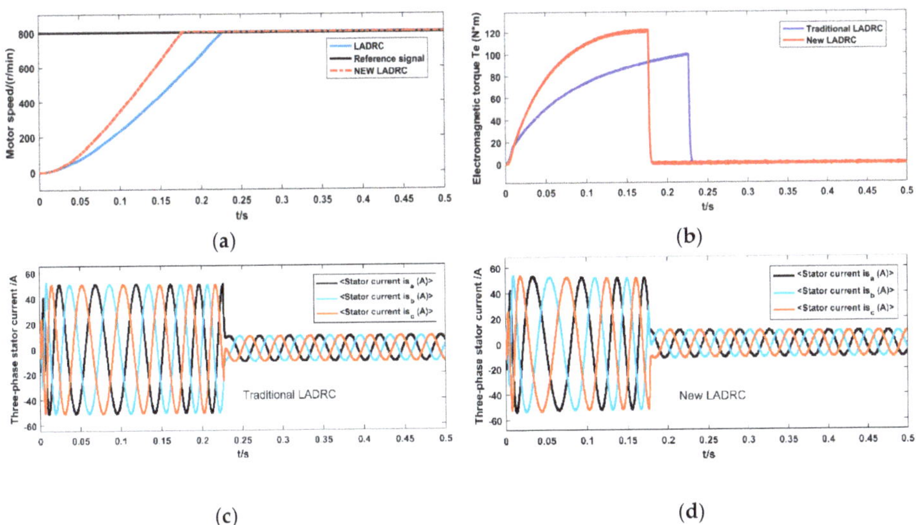

Figure 15. High-speed simulation of the asynchronous motor.

4.2. Steady State Error of Asynchronous Motor Controller at Different Given Speeds

Figure 16a is the local amplification of Figure 15a. Although the steady-state error of the system under the action of the traditional Linear Active Disturbance Rejection Controller is not large at a given reference speed of 200 r/min, the steady-state error of the asynchronous motor under the action of the improved Linear Active Disturbance Rejection Controller is smaller. Figure 16b is the local amplification of Figure 16a. Compared with the given reference speed of 200 r/min, the steady-state error of the asynchronous motor under the action of the improved Linear Active Disturbance Rejection Control is still better than the traditional Linear Active Disturbance Rejection Controller, which, although, becomes larger under the given reference speed of 200 r/min.

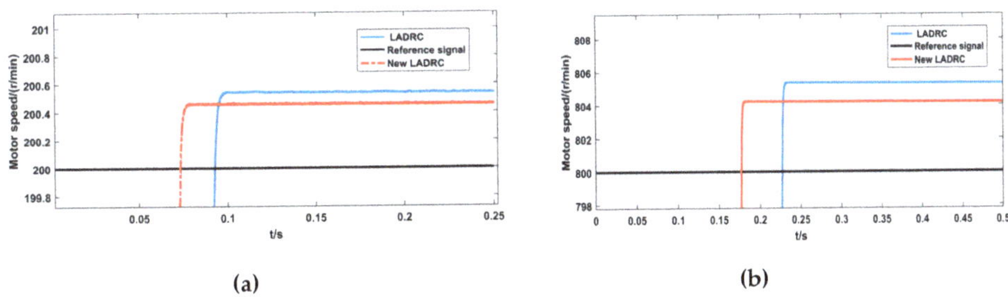

Figure 16. Steady-state error of the system at different speeds.

PI control uses error feedback to eliminate the error. When the system output is equal to the given input, the controller output is zero. At this time, the motor cannot maintain the current speed and deviates from the given value, resulting in the steady-state error of motor speed. The coupling between the excitation subsystem and the torque subsystem increases with the increase of the speed, resulting in the steady-state error of the speed of the motor in Figure 16b at high speed higher than that in Figure 16a at low speed.

4.3. Immunity Performance of the Asynchronous Motor Controller at Different Given Speeds

Figure 17a is a simulation diagram of the motor speed during the process of a sudden increase in the mechanical torque by 10 N·m in 0.35 s and a subsequent sudden decrease in the mechanical torque by 10 N·m in 0.4 s at a given speed of 200 rev/min (returning to the state before the system loading). Figure 17b is a simulation diagram of the motor speed during the process of a sudden increase in the mechanical torque by 10 N·m in 0.35 s and subsequent sudden decrease in the mechanical torque by 10 N·m in 0.4 s (returning to the state before the system loading) at a given speed of 800 rev/min. From Figure 17a,b, it can be seen that the motor speed controlled by the traditional LADRC fluctuated more after the mechanical load of the motor suddenly increased; therefore, the immunity of the improved LADRC was better than that of the traditional LADRC.

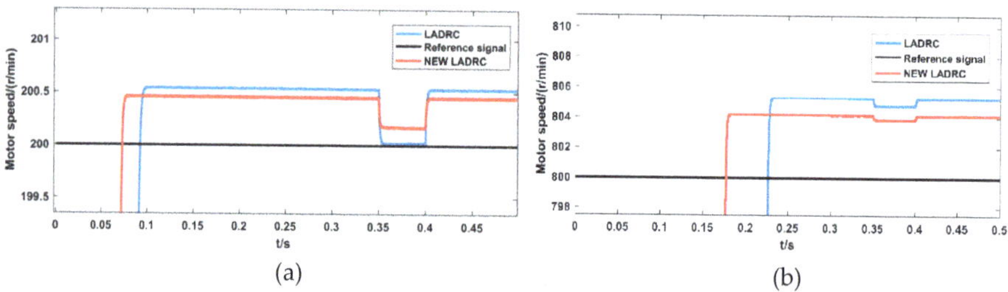

(a)   (b)

Figure 17. Sudden load increase and decrease of the motor at different speeds.

The coupling between the excitation subsystem and the torque subsystem increases to the increase in speed, resulting in the steady-state error of the motor in Figure 17b at a high speed greater than that in Figure 17a at low speed. The improved linear expansion observer can estimate the total disturbance more accurately, reduce the coupling between the two subsystems to a certain extent, and improve the control of electromagnetic torque. Since the improved Linear Active Disturbance Rejection Controller is applied to the current inner loop decoupling and the outer loop still adopts PI control, the improvement of load torque and the steady-state error is not obvious.

5. Conclusions

The key to the performance of an ADRC is whether the extended state observer can accurately estimate the state variables of the system. There is a cross-coupling term in the state equation of the asynchronous motor in the synchronous rotating coordinate system. The control of the torque component and the excitation component of the motor stator current will affect each other, thus further affecting the dynamic and static performances of the system.

Actual systems always have unknown dynamic characteristics, i.e., the uncertainty of the model. In the control system, there are often various external disturbances, such as the control quantity disturbance or the measurement noise. LESOo regards the coupling of the dynamic model of the asynchronous motor system as a part of the total disturbance of the system. The disturbance is added to the input of the system model using feedforward compensation. After compensation, the model of the controlled asynchronous motor is transformed into the integrator series type.

The key to the motor speed regulation lies in the regulation of the electromagnetic torque. The improved LESOo had a higher observation accuracy, the estimated value of total disturbance was closer to the real value, the coupling degree between the torque component and excitation component was smaller after compensation, and the improved LESOo could control the torque component more independently; therefore, the control effect of the motor speed was better. Theoretical analysis and simulation results showed that the control effect of the improved LADRC was better than that of the traditional LADRC.

Author Contributions: C.W. was responsible for the theoretical analysis and simulation, and wrote the first draft of the paper. X.Z. and Y.M. provided theoretical guidance and content revision. All authors have read and agreed to the published version of the manuscript.

Funding: This work was funded by the National Natural Science Foundation of China (no. 51877152) and the Natural Science Foundation of Tianjin of China (no. 18JCZDJC97300).

Acknowledgments: Thanks to X.Z. and Y.M. for their theoretical guidance during the writing of this paper.

Conflicts of Interest: The authors declare no conflict of interest.

Abbreviations

Acronym	Definition
LESO	Linear extended state observer
LSEF	Linear state error feedback
LTD	Linear tracking differentiator
LADRC	Linear active disturbance rejection control

Appendix A

Table A1. Three-phase cage asynchronous motor parameters.

Parameter	Symbol	Value	Unit
Rated Capacity	S_N	3730	VA
Rated voltage	U_N	220	V
Rated frequency	f	50	Hz
Stator resistance	R_s	0.435	Ω
Rotor resistance	R_r	0.069	Ω
Stator inductance	Ls	0.079	H
Rotor inductance	Lr	0.071	H
Stator and rotor mutual inductance	Lm	0.069	H
Pole pairs of asynchronous motor	n_p	2	

References

1. Fedor, P.; Perdukova, D.; Kyslan, K.; Fedak, V. *Stable and Robust Controller for Induction Motor Drive*; IEEE: Piscataway, NY, USA, 2018; pp. 764–769.
2. Hannan, M.A.; Ali, J.A.; Mohamed, A.; Hussain, A. Optimization techniques to enhance the performance of induction motor drives: A review. *Renew. Sustain. Energy Rev.* **2018**, *81*, 1611–1626. [CrossRef]
3. Happyanto, D.C.; Fauzi, R.; Hair, J. *Backstepping Development as Controller in Fast Response Three Phase Induction Motor Based on Indirect Field Oriented Control*; IEEE: Piscataway, NY, USA, 2016; pp. 25–30.
4. Mehazzem, F.; Nemmour, A.L.; Reama, A. Real time implementation of backstepping-multiscalar control to induction motor fed by voltage source inverter. *Int. J. Hydrogen Energy* **2017**, *42*, 17965–17975. [CrossRef]
5. Feng, G.; Liu, Y.F.; Huang, L.P. A new robust algorithm to improve the dynamic performance on the speed control of induction motor drive. *IEEE Trans. Power Electron.* **2004**, *19*, 1614–1627. [CrossRef]
6. Ansari, M.N.; Dalal, A.; Kumar, P. A Method for Determining Nonlinear Inductances of Electrical Equivalent Circuit for Three-Phase Induction Motor. *Electr. Power Compon. Syst.* **2018**, *46*, 379–390. [CrossRef]
7. Aydemir, M.; Arikan, K.B. Evaluation of the Disturbance Rejection Performance of an Aerial Manipulator. *J. Intell. Robot. Syst.* **2020**, *97*, 451–469. [CrossRef]
8. Sira-Ramirez, H.; Linares-Flores, J.; Garcia-Rodriguez, C.; Antonio Contreras-Ordaz, M. On the Control of the Permanent Magnet Synchronous Motor: An Active Disturbance Rejection Control Approach. *IEEE Trans. Control Syst. Technol.* **2014**, *22*, 2056–2063. [CrossRef]
9. Ahmad, S.; Ali, A. Active disturbance rejection control of DC-DC boost converter: A review with modifications for improved performance. *IET Power Electron.* **2019**, *12*, 2095–2107. [CrossRef]
10. Abdul-Adheem, W.R.; Ibraheem, I.K. Decoupled control scheme for output tracking of a general industrial nonlinear MIMO system using improved active disturbance rejection scheme. *Alex. Eng. J.* **2019**, *58*, 1145–1156. [CrossRef]

11. Han, J.Q.; Wang, X.J. Time scale and nonlinear PID controller of the system. In Proceedings of the 1994 China Control Conference, Taiyuan, Shanxi, China, 1 August 1994; p. 8.

12. Hua, X.X.; Huang, D.; Guo, S.H. Extended State Observer Based on ADRC of Linear System with Incipient Fault. *Int. J. Control Autom. Syst.* **2019**. [CrossRef]

13. Laghridat, H.; Essadki, A.; Nasser, T. Comparative Analysis between PI and Linear-ADRC Control of a Grid Connected Variable Speed Wind Energy Conversion System Based on a Squirrel Cage Induction Generator. *Math. Probl. Eng.* **2019**. [CrossRef]

14. Lin, X.; Xiahou, K.S.; Liu, Y.; Zhang, Y.B.; Wu, Q.H. *Maximum Power Point Tracking of DFIG-WT Using Feedback Linearization Control Based Current Regulators*; IEEE: Piscataway, NY, USA, 2016; pp. 718–723.

15. Li, H.; Li, S.; Lu, J.; Qu, Y.; Guo, C. A Novel Strategy Based on Linear Active Disturbance Rejection Control for Harmonic Detection and Compensation in Low Voltage AC Microgrid. *Energies* **2019**, *12*, 3982. [CrossRef]

16. Bose, S.; Hote, Y.V.; Siddhartha, V. *Analysis and Application of Linear ADRC for the Control of DC-DC Converters*; IEEE: Piscataway, NY, USA, 2019; pp. 436–441.

17. Gu, J.; AI, Y.; Shan, X.; Wang, Z.; Liu, M.; Xiong, Z. Improvement of linear extended state observer and its application in coarse tracking of space optical communication. *Infrared Laser Eng.* **2016**, *45*. [CrossRef]

18. He, H.C.; Sun, L.; Zhang, Y.; Zhu, Q. Active disturbance rejection control of induction motor based on vector control. *J. Electr. Mach. Control* **2019**, *23*, 120–125.

19. Zhou, Y.S. A survey of decoupling control strategy for asynchronous motor. *Small Medium Mot.* **2005**, *2*, 56, 60–64.

20. Ren, Y.; Zhao, G.H.; Liu, H. Design of fast arctangent tracking differentiator. *Control Eng.* **2019**, *26*, 412–416.

21. Tan, S.L.; Lei, H.m.; Wang, P.F. Inverse control of hypersonic vehicle with tracking differentiator. *Acta Astronaut.* **2019**, *40*, 673–683.

22. Shen, Z.P.; Cao, X.M. An extended observer based output feedback control for trajectory tracking dynamic surface of an input constrained four rotor vehicle. *Syst. Eng. Electron. Technol.* **2018**, *40*, 2766–2774.

23. Yu, B.; Liu, Y.L.; Gao, Z.J.; Kong, X. Dong Position disturbance rejection control of hydraulic drive unit based on extended observer. *Hydraul. Pneum.* **2018**, *6*, 1–8.

24. Zhou, X.S.; Tian, C.W.; Ma, Y.J.; Liu, S.J.; Zhao, J.; Liu, J. SHAPF model and current tracking control strategy based on LADRC. *Power Autom. Equip.* **2013**, *33*, 49–54.

25. Sun, D.S. SRM control system based on time-varying parameter Active disturbance rejection. *Electr. Drive* **2018**, *48*, 19–22.

26. Yang, L.; Zeng, J.; Ma, W.J.; Huang, Z.L. Voltage control of micro grid inverter based on improved second-order linear auto disturbance rejection technology. *Power Syst. Autom.* **2019**, *43*, 146–158.

Article

Optimal Site Selection for a Solar Power Plant in the Mekong Delta Region of Vietnam

Chia-Nan Wang [1,*], Van Tran Hoang Viet [1,*], Thanh Phong Ho [2,*], Van Thanh Nguyen [2] and Syed Tam Husain [2]

[1] Department of Industrial Engineering and Management, National Kaohsiung University of Science and Technology, Kaohsiung 80778, Taiwan
[2] Department of Logistics and Supply Chain Management, Hong Bang International University, Ho Chi Minh 723000, Vietnam; thanhnv@hiu.vn (V.T.N.); husains@hiu.vn (S.T.H.)
*   Correspondence: cn.wang@nkust.edu.tw (C.-N.W.); vanviet@tnut.edu.vn (V.T.H.V.); phonght@hiu.vn (T.P.H.)

Received: 21 May 2020; Accepted: 30 July 2020; Published: 6 August 2020

Abstract: Following the recent development trend in the struggle for cleaning the earth's environment, solar is the one of most promising area that can partially be used as a replaceable energy from non-renewable fuel sources. As such, it plays a significant role in protecting the environment from global warming. As solar power does not emit harmful gases into the atmosphere, its production, distribution, setup, and operation are vital should the production remain constant. Even solar energy waste emissions are small; when compared to current energy sources, the amount of harmful gases is negligible. This paper presented an integrated approach for site of solar plants by using data envelopment analysis (DEA) and Fuzzy Analytical Network Process (FANP). Furthermore, these integrated methodologies, incorporated with the most relevant parameters of requirements for solar plants, are introduced. First, the paper considers an integrated hierarchical DEA and FANP model for the optimal geographical location of solar plants in Mekong Delta Region, Vietnam. Using the proposed model for implementation would allow the renewable energy policy makers to select and control the optimal location for allocating and constructing a solar energy power plant in Vietnam. This is the preferred strategy for location optimization problems associated with solar plant units in Vietnam and around the world.

Keywords: renewable energy; solar power plant; Data Envelopment Analysis (DEA); Fuzzy Analytical Network Process (FANP); Fuzzy Theory

---

1. Introduction

Energy has been a fundamental driver of global economies in order maintain development in almost any industry. It is compulsory that the sources of energy must be maintained and continue developing to meet the demand. With fossil fuel continuing to be proven to be detrimental to the environment, solar energy is continuing to become a modern, growing, and potential source of renewable energy. According to the International Energy Agency (IEA) in 2015, the consumption of fossil fuels coming from coal, oil, and natural gas accounts for 81.6% of the total energy consumption amongst the 38 leading countries. From the same data, wind, and solar energy only accounts for 1.5% of the energy consumption [1]. As also shown in Figure 1, solar energy has the highest development compared to other renewable energies such as hydropower, onshore and offshore wind energy, and bioenergy [2]. According to NASA's physics and astronomy calculations, solar energy is still available for about 6.5 billion years before the sun fades away, which means that the energy source can be utilized for a longer and more durable time before it is projected to deplete [3].

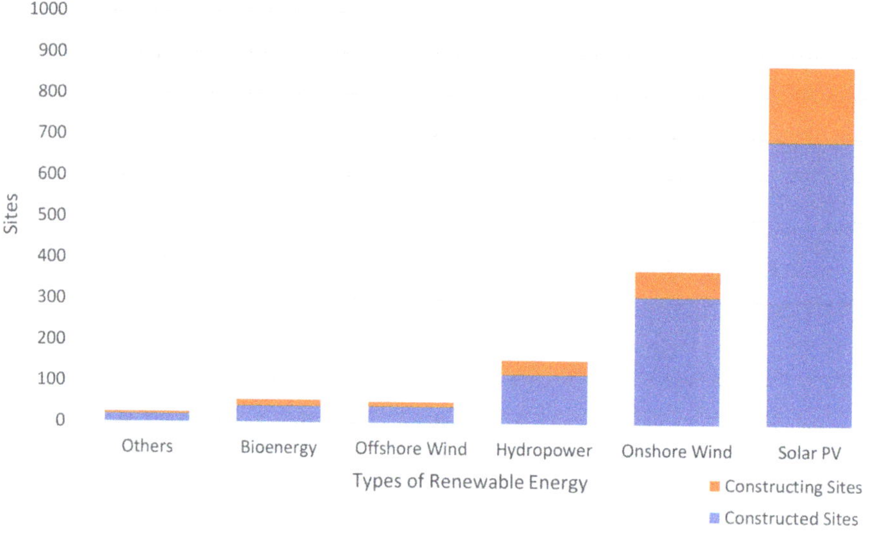

Figure 1. Comparison of different renewable energy sources [4].

Solar energy takes the direct sun rays from the sun and transforms them into usable energy consumed for many purposes. The major limitations of solar energy include geographical conditions, high dependency on technology development, large installation area with a high initial cost. Additionally, with such limitations come additional, big environmental changes where modifications could leave an impact on local ecosystems should the development a plant be initialized with the trend of sustainable development, and maintaining the balance of the triple bottom line of environmental, social, and economical elements must be considered [5].

Solar energy is still not a popular energy source amongst the economic and social development of the provinces in the Mekong Delta region, Vietnam. However, according to Vietnam Academy of Science and Technology (VAST), Vietnam is one of the potential countries for solar energy development when looking at the data for the amount of sun rays the world receives annually, shown in Figure 2 with a second highest overall evaluation of 4–5 Kwh/m^2/day [6].

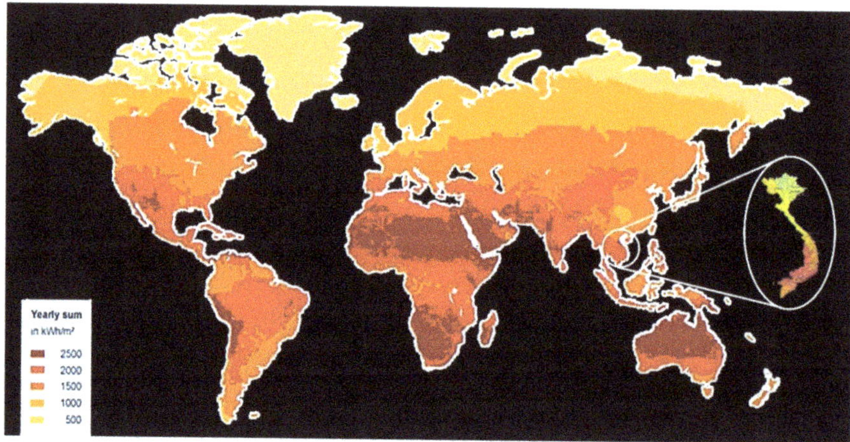

Figure 2. Average amount of potential solar energy output [6].

In fact, if more than 50% [7] of the energy supply for a national electricity system is imported from other countries, the risk of energy unavailability will be extremely high. Once that supply is depleted or affected by natural disasters, accidents, armed conflicts, sabotage, policy changes or strikes within coal mines, energy security will be directly affected. Besides, the coal transportation from a remote area to the electric production factory, which is done mainly by sea, also has many potential implications for the environment, social security, and economy. The pollution of dust, noise, and ash accumulation from the operation of coal-fired power plants can cause frustration in the local community and sow the seeds for protests, creating long-term instability in terms of social security [8]. Located in the tropical and monsoon region, with an abundant solar energy, the Mekong Delta of Vietnam is entering a period of strong solar power development. Solar energy has been selected as one of the major alternative renewable energy sources in the energy development strategy of the Mekong Delta region. Solar Radiation Maps of Vietnam is shown in Figure 3.

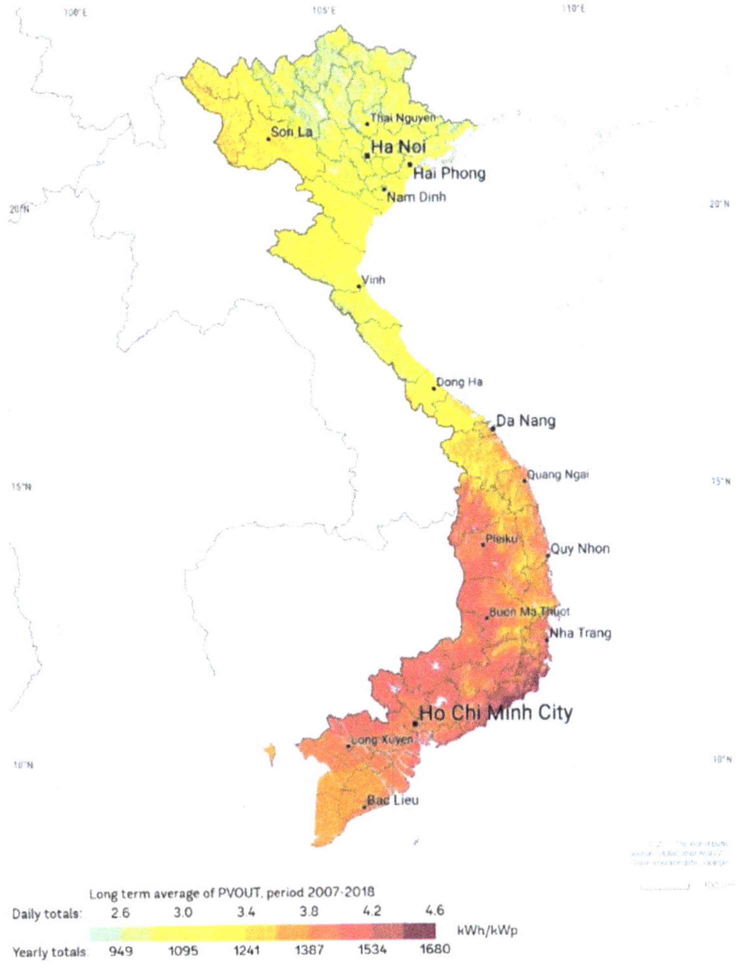

Figure 3. Solar Radiation Maps of Vietnam [9].

Currently, there is very little research for developing a solar power plant in Vietnamese Mekong Delta region. Determining an optimal location, which is sustainable and satisfies the triple bottom line concepts, is essential for the growth of the country's power supply and alleviates the environmental damage from fossil fuel consumption by promoting a renewable alternative energy supply. Therefore, this problem is tackled as a Multi-Criteria Decision Making (MCDM) problem that requires both quantitative and qualitative information to be considered. The MCDM problem is commonly defined as determining an optimal decision based on a finite set of available decision alternatives with multiple, and potentially conflicting, criteria.

As mentioned, the purpose of this paper is now to propose a suitable MCDM methodology for localizing and selecting an optimal area in Vietnam in developing a solar power plant. In the next section, the literature will be presented to discuss the first stage of the paper. Discussions and results' analysis are explained towards the end of the paper.

2. Literature Review

Nowadays, there are many researchers applied MCDM models to any fields of sciences. A. Kengpol et al. [10] initially conducted a study on proposing a solar power plant location to avoid flooding in Thailand using geographical information system (GIS) to determine an optimum site condition. Upon the study, a number of geographical conditions had to be considered which developed into an MCDM problem of ranking the priorities amongst the conditions using Fuzzy Analytic Hierarchy Process (FAHP) and the Technique of Order Preference by Similarity to Ideal Solution (TOPSIS).

A. Azadeh et al. [11] integrated a hierarchical model for solar plants' location selection by data envelopment analysis (DEA), principal component analysis (PCA) and numerical taxonomy (NT). Anuchit Thongpun et al. [12] focused on a building location selection approach for locating a solar power plant in Thailand based on DEA models. A. Azadeh et al. [13] applied an artificial neural network (ANN) and fuzzy data envelopment analysis (FDEA) for location optimization of solar plants with uncertainty and complexity.

Amy H. I. Lee et al. [14] integrated MCDM model to set the assurance region (AR) of the quantitative factors, and the AR is incorporated into data envelopment analysis (DEA), additionally adopting a fuzzy analytic hierarchy process (FAHP) for the location of a PV solar plant. Adnan Sozen et al. [15] presented an approach for the location of solar plants by data envelopment analysis (DEA). Ali Azadeh et al. [16] presented an integrated fuzzy DEA model for decision making on wind plant locations. Shinya Yokota et al. [17] applied data envelopment analysis (DEA) for the optimal allocation of mega-solar.

Chabuk et al. [18] discussed two MCDM methods to propose two alternative landfill sites for each area in the studied location, which were AHP and Ratio Scale Weighting (RSW), using multiple environmental factors as criteria for evaluation. The methods successfully proposed a geographical map that addressed the suitable locations throughout the studied sector. Chakraborty et al. [19] also proposed multiple MCDM methods including AHP for assigning criteria weights; Grey Rational Analysis (GRA), Multi Objective Optimization on the Basis of Ratio Analysis (MOORA), Elimination of Choice Translating Reality (ELECTRE II), and Operational Competitiveness Rating Analysis (OCRA) for ranking the results of each location alternative; and Spearman's rank correlation coefficient, Kendall's rank correlation coefficient, agreement between the top three ranked alternatives were used to compare each ranking methodology with another; finally, the REGIME was used for the final evaluation of each methodology after all the applied evaluating methodologies were calculated. With a huge comparison study between the methods, the author was able to utilize which methodology was most suitable for choosing the most suitable supplier.

Amy H. I. Lee et al. [20] proposed MCDM model to decide the most suitable photovoltaic solar plant allocation by using the interpretive structural modeling (ISM), the Serbian VlseKriterijumska OptimizacijaI Kompromisno Resenje (VIKOR), meaning multi-criteria optimization with a compromise solution, fuzzy analytic network process (FANP). A. Azadeh et al. [21] integrated hierarchical Data Envelopment Analysis for the location optimization of wind plants in Iran. This model would enable the energy policy makers to select the best possible location for construction of a wind power plant with the lowest possible cost. Lei Fang et al. [22] applied the DEA model and goal programming to evaluate the relative efficiency of each potential location.

Chia-Nan Wang et al. [9] proposed a similar MCDM approach to determine a solar power plant location for the entire country of Vietnam using the methods of DEA, FAHP, and TOPSIS to evaluate qualitative and quantitative criteria. Ehsan Dehghani et al. [23] evaluated different areas for solar plants according to a set of social, geographical, and technical criteria through a data envelopment analysis (DEA) model. In this study, the DEA model considers both information of the efficient and anti-efficient frontiers to raise discrimination power in DEA analysis. Ali Mostafaeipour et al. [24] applied Data Envelopment Analysis (DEA) methodology to prioritize cities for installing the solar-hydrogen power plant so that one candidate location was selected for each city. A. Azadeh et al. [25] presented a technical and economic research for allocation of solar plants by using multivariable methods namely, Data Envelopment Analysis (DEA) and Principle Component Analysis (PCA). A hybrid model for the allocation of solar plants was presented by the utilization of most related parameters to solar plants and an integrated DEA-PCA approach. Seong Kon Lee et al. [26] proposed a hybrid model including a fuzzy Analytic Hierarchy Process (AHP)/Data Envelopment Analysis (DEA) for efficiently allocating energy R&D resources in the case of energy technologies against high oil prices.

As the literature review shows, the amount of studies that apply the MCDM approach to various fields of science and engineering has increased in number in recent years. Location selection is one of the fields where the MCDM model has been employed, especially in the renewable energy sector, where decision makers must evaluate both qualitative and quantitative factors. Although some studies have reviewed the applications of MCDM approaches in solar power plant location selection, very few works have focused on this problem in a fuzzy environment. This is a reason why we proposed a fuzzy MCDM model in this study.

3. Methodology

3.1. Basic Theory

3.1.1. Fuzzy Set

A fuzzy set is used to address the issues that exist in an uncertain environment. It was proposed by Zadeh when attempting to solve multiple criteria decision-making problems. If the set $\tilde{Z}$ is a triangular fuzzy number (TFN), each value of the membership function is in the range [0, 1]:

$$\mu_{\tilde{Z}}(x) = \begin{cases} \frac{(x-j)}{(k-j)} & j \leq x \leq k \\ \frac{(l-x)}{(l-k)} & k \leq x \leq l \\ 0 \end{cases} \tag{1}$$

Each value of membership, which includes the left (the lowest value to evaluate) and right (the highest value to evaluate) representatives, of a TFN is shown below:

$$\tilde{N} = (N^{l(y)}, N^{r(y)}) = (j - (k - j)\, y,\ l + (k - d)\, y),\ y \in [0, 1].$$

A triangular fuzzy number can be described as in Figure 4.

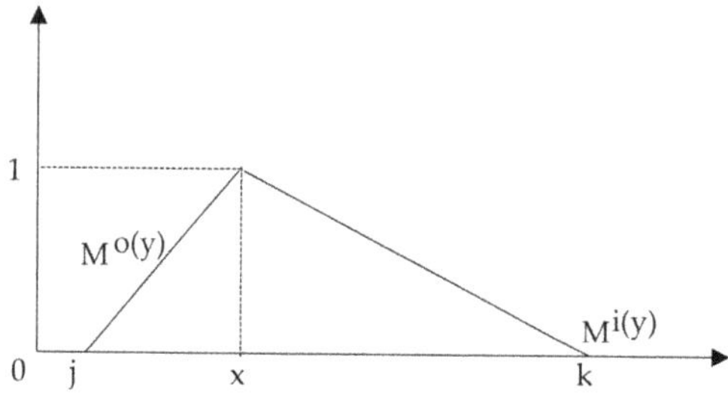

Figure 4. A triangular fuzzy number (TFN).

3.1.2. Fuzzy Analytic Network Process

Different to the AHP model, a strict hierarchical structure is not required for the ANP model. ANP models allow control elements, that can be controlled by attribute clusters or difference levels. When the elements are at the same level, they can present multiple control factors. A systematic approach in the interaction or feedback between factors is explained using the degree of interaction and interdependence between factors.

Using the rating scale of expert as the basis and data, the decision maker will pair each factor where the weight of each factor is matrixed and determined for the ANP model.

The AHP model then uses a quantitative pair comparison with a priority ratio of 1–9 to set the priority level for each level of the system, where 1 is the lowest priority and 9 is the highest priority. Meanwhile, the ANP model allows making complex relationships between criteria and their rank in the system. The 1–9 scale of the AHP model is shown in Table 1.

Table 1. The 1–9 scale [6].

Importance Intensity	Definition
1.	Equal significance
3.	Moderate significance
5.	Strongly higher significance
7.	Very strong higher significance
9.	Extremely high significance
2, 4, 6, 8.	Intermediate values

The display and goals in the pairwise comparison process, which were considered disadvantages of ANP, were overcome with the development of the FANP model. The FANP model uses a set of values to incorporate decision makers in an uncertain environment, while a crisp value is presented in the ANP model. The Saaty's model is used to convert the values for the fuzzy prioritization model to easily fix the conversion values, where $O_{ab} = (O_{ab}^x, O_{ab}^o, O_{ab}^v)$ is a TFN with the core O_{ab}^o, the support $[O_{ab}^x, O_{ab}^v]$, and the TFN, as shown in Figure 5.

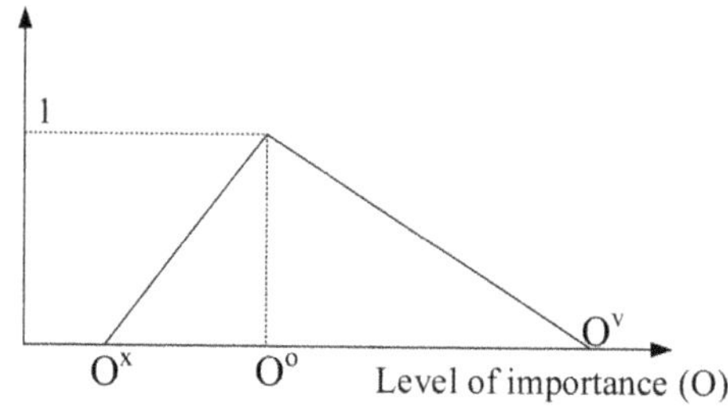

Figure 5. A triangular fuzzy number (TFN).

Table 2 presents the 1–9 fuzzy conversion scale.

Table 2. The 1–9 fuzzy conversion scale.

Importance Intensity.	Triangular. Fuzzy Scale.
1.	(1.0, 1.0, 1.0).
2.	(1.0, 1.0, 2.0).
3.	(1.0, 2.0, 3.0).
4.	(2.0, 3.0, 4.0).
5.	(3.0, 4.0, 5.0).
6.	(4.0, 5.0, 6.0).
7	(5.0, 6.0, 7.0).
8	(7.0, 8.0, 9.0).
9	(9.0, 9.0, 9.0).

At the reverse level to O_{ab}, expressing the non-preference is also shown by a TFN: $(1/O_{ab}^v, 1/O_{ab}^o, 1/O_{ab}^x)$. Using the fuzzy Saaty's matrix, the weights of the criteria can be determined into four steps that are used to input the data:

1. Using Equations (2)–(4), the fuzzy synthetic extensions $K_a(k_a^x, k_a^o, k_a^v)$ calculation can be transformed into TNT.

$$K_a = \sum_{b=1}^{n} O_{ab} \otimes \left(\sum_{a=1}^{n} \sum_{b=1}^{n} O_{ab} \right)^{-1} \tag{2}$$

$$\sum_{j=1}^{n} O_{ab} = \left(\sum_{b=1}^{n} O_{ab}^x, \sum_{b=1}^{n} O_{ab}^o, \sum_{b=1}^{n} O_{ab}^v \right) \tag{3}$$

$$O_{ab}^{-1} = 1/O_{ab}^v, 1/O_{ab}^o, 1/O_{ab}^x \tag{4}$$

$$O \otimes N = (O_x.N_x,\ O_0.N_0,\ O_v.N_v) \tag{5}$$

Let $a = 1, 2, \ldots, n$, in which a and b specifically are TFN (O_x, O_o, O_v) and (N_x, N_0, N_v) where x is the minimum value, o is the average value, and v is the maximum value;

2. Using relationships with the fuzzy valued for addressing the weight of criteria. In order to determine certain fuzzy extensions, they are calculated by using the minimum fuzzy extension of valued relation $\leq$ at [5] with weights Q_a calculated as follows

$$Q_a = min_b \left\{ \frac{k_b^v - k_a^v}{\left(k_a^o - k_a^v \right) - \left(k_b^o - k_b^x \right)} \right\} \tag{6}$$

where $a, b = 1, 2, \ldots ., n$;

3. Standardize the weights. If the decision maker expects to obtain the total weights in one matrix equal to 1, final weights q_i are solved by (7)

$$q_a = Q_a / \sum_{a=1}^{n} Q_a \tag{7}$$

where $a, b = 1, 2, \ldots , n$;

4. An evaluation of a Saaty's matrix is used to test for its consistency. The matrix of the weights and criteria are consistent and sufficient if inequality of the Consistency Ratio (CR) from Equation (8) is defined as follows using the Consistency Index (CI) and Random Index (RI):

$$CR = \frac{CI}{RI} = \frac{\bar{\lambda} - n}{(n-1) \times RI} \leq 0.1 \tag{8}$$

3.2. Data Envelopment Analysis Model

3.2.1. Charnes–Cooper–Rhodes model (CCR model)

The CCR model, as the fundamental model for the DEA model, is defined as follows:

$$\max_{c.a} \xi = \frac{a^Z y_0}{c^Z x_0}$$

S.t :

$$a^Z y_e - c^Z x_e \leq 0, \ e = 1, \ 2, \ldots, n \tag{9}$$
$$a \geq 0$$
$$c \geq 0$$

The defined constraints ensure that the ratio of virtual output to virtual input cannot exceed 1 per decision making unit (DMU). The objective is to obtain a rate of weighted output for every weighted inputs. Subject to the constraints, the optimal goal value ξ^* can only reach a maximum of 1.

DMU_0 is CCR's efficient if $\xi^* = 1$. The result must include a minimum of 1 optima $a^* > 0$ and $c^* > 0$. In addition, the fractional program can be defined as a linear programming problem (LP) as follows:

$$\max_{c.a} \xi = a^Z y_0$$

S.t :

$$c^Z x_0 - 1 = 0$$
$$a^Z y_e - c^Z x_e \leq 0, \ e = 1, \ 2, \ldots, n \tag{10}$$
$$c \geq 0$$
$$a \geq 0$$

The linear program (10) provides an equal result to the fractional program (9). The linear program from the Farrell model (10) has a variable ξ and a nonnegative vector $\alpha = \alpha_1, \alpha_2, \alpha_3, \ldots, \alpha_f$ as:

$$\max \sum_{d=1}^{m} s_i^- + \sum_{g=1}^{o} s_g^+$$

S.t :

$$\sum_{e=1}^{n} x_{de}\alpha_d + s_d^- = \xi x_{d0}, \ b = 1, \ 2, \ldots, p$$
$$\sum_{e=1}^{n} y_{ge}\alpha_e - s_g^+ = y_{g0}, \ g = 1, \ 2, \ldots, o \tag{11}$$
$$\alpha_e \geq 0, \ e = 1, \ 2, \ldots, \ n$$
$$s_d^- \geq 0, \ d = 1, \ 2, \ldots, \ p$$
$$s_g^+ \geq 0, \ g = 1, \ 2, \ldots, \ o$$

The model (3) provides a feasible solution, $\xi = 1, \alpha_0^* = 1, \alpha_j^* = 0, (j \neq 0)$, in which the optimal solution is affected when ξ^* is not higher than 1. A specific DMU is provided when the optimal solution, ξ^*, is calculated. For each DMU_e, the process will repeat for every $e = 1, 2, \ldots, n$. When $\xi^* < 1$, the DMUs are inefficient. If $\xi^* = 1$, the DMUs then are classified as boundary units. By invoking a linear program, we can prevent weakly efficient points d as follows:

$$max \sum_{d=1}^{m} s_d^- + \sum_{g=1}^{s} s_g^+$$

S.t :

$$\begin{aligned}
&\sum_{e=1}^{n} x_{de}\alpha_e + s_d^- = \xi x_{d0}, \; d = 1, 2, \ldots, p \\
&\sum_{e=1}^{n} y_{ge}\alpha_e - s_g^+ = y_{r0}, \; g = 1, 2, \ldots, o \\
&\alpha_e \geq 0, \; e = 1, 2, \ldots, n \\
&s_d^- \geq 0, \; d = 1, 2, \ldots, p \\
&s_g^+ \geq 0, \; g = 1, 2, \ldots, o
\end{aligned} \tag{12}$$

For this situation, we clarify that the optimal solution, ξ^*, is not affected by the results from s_d^- and s_g^+.

For both (1) $\xi = 1$ and (2) $s_d^{-*} = s_g^+ = 0$, DMU_0 achieves 100% accuracy and efficiency. For both (1) $\xi^* = 1$ and (2) $s_d^{-*} \neq 0$ and $s_g^+ \neq 0$ for d or g in optimal options, the performance of DMU_0 is weakly efficient. Thus, following the development procedure to solve the problem is as follows:

$$min\theta - \mu\left(\sum_{d=1}^{m} s_d^- + \sum_{g=1}^{s} s_g^+\right)$$

S.t :

$$\begin{aligned}
&\sum_{b=1}^{n} x_{de}\alpha_e + s_d^- = \xi x_{d0}, \; d = 1, 2, \ldots, p \\
&\sum_{e=1}^{n} y_{ge}\alpha_e - s_g^+ = y_{g0}, \; g = 1, 2, \ldots, o \\
&\alpha_e \geq 0, \; e = 1, 2, \ldots, n \\
&s_d^- \geq 0, \; d = 1, 2, \ldots, p \\
&s_g^+ \geq 0, \; g = 1, 2, \ldots, o
\end{aligned} \tag{13}$$

In this case, s_b^- and s_r^+ variables are first used to transform the inequalities into equivalent equations. For (13), it is the same in terms of methods as when we solve (3) by minimizing ξ in the first stage and then fixing $\xi = \xi^*$ as in (4), where the slacks' variables provide the highest values but the previously determined value of $\xi = \xi^*$ is not affected. The objective is converted from maximum to minimum, as in (9), to obtain the following:

$$\max_{c.a} \xi = \frac{c^Z x_0}{a^Z y_e}$$

S.t :

$$\begin{aligned}
&a^Z x_0 \leq c^Z y_e, \; e = 1, 2, \ldots, n \\
&c \geq \varepsilon > 0 \\
&a \geq \varepsilon > 0
\end{aligned} \tag{14}$$

If the non-Archimedean value and the $\varepsilon > 0$ are displayed, the input models are similar to models (10) and (13) as follows:

$$\max_{c.a} \xi = c^Z x_0$$

S.t :

$$\begin{aligned}
&a^Z y_0 = 1 \\
&c^Z x_0 - a^Z y_e \geq 0, \; e = 1, 2, \ldots, n \\
&c \geq \varepsilon > 0 \\
&a \geq \varepsilon > 0
\end{aligned} \tag{15}$$

and:

$$max\phi - \varepsilon\left(\sum_{d=1}^{m} s_i^- + \sum_{g=1}^{s} s_g^+\right)$$

S.t :

$$\sum_{e=1}^{n} x_{de}\alpha_e + s_d^- = x_{d0}, \; d = 1, 2, \ldots, p$$
$$\sum_{e=1}^{n} y_{ge}\alpha_e - s_g^+ = \varnothing y_{g0}, \; g = 1, 2, \ldots, q$$
$$\alpha_e \geq 0, \; e = 1, 2, \ldots, n$$
$$s_d^- \geq 0, \; d = 1, 2, \ldots, p$$
$$s_g^+ \geq 0, \; g = 1, 2, \ldots, o$$

(16)

A dual multiplier model of the CCR input-oriented (CCR-I) is expressed as:

$$maxz = \sum_{g=1}^{q} \partial_g y_{g0}$$

S.t :

$$\sum_{g=1}^{o} \partial_g y_{re} - \sum_{g=1}^{o} a_g y_{ge} \leq 0$$
$$\sum_{d=1}^{p} a_d x_{d0} = 1$$
$$c_g, a_d \geq \varepsilon > 0$$

(17)

A dual multiplier model of the CCR output-oriented (CCR-O) is also expressed as:

$$mino = \sum_{d=1}^{p} a_d x_{d0}$$

S.t :

$$\sum_{d=1}^{p} a_d x_{de} - \sum_{g=1}^{o} \partial_g y_{ge} \leq 0$$
$$\sum_{g=1}^{o} \partial_g y_{g0} = 1$$
$$c_g, a_d \geq \varepsilon > 0$$

(18)

3.2.2. Banker Charnes Cooper Model (BCC Model)

The input-oriented BBC model (BCC-I) was introduced by Banker et al., in which the efficiency of DMU_0 is assessed by solving the following LP (19):

$$\xi_B = min\xi$$

S.t :

$$\sum_{e=1}^{n} x_{de}\alpha_e + s_d^- = \xi x_{d0}, \; d = 1, 2, \ldots, p$$
$$\sum_{e=1}^{n} y_{ge}\alpha_e - s_g^+ = y_{g0}, \; g = 1, 2, \ldots, o$$
$$\sum_{k=1}^{n} \alpha_k = 1$$
$$\alpha_k \geq 0, \; k = 1, 2, \ldots, n$$

(19)

By invoking a linear program, we can prevent the weakly efficient points as in the following:

$$max \sum_{d=1}^{m} s_d^- + \sum_{g=1}^{s} s_g^+$$

S.t :

$$\sum_{e=1}^{n} x_{de}\alpha_e + s_d^- = \xi x_{d0}, \; d = 1, 2, \ldots, p$$
$$\sum_{e=1}^{n} y_{ge}\alpha_e - s_g^+ = y_{g0}, \; g = 1, 2, \ldots, o \tag{20}$$
$$\sum_{k=1}^{n} \alpha_k = 1$$
$$\alpha_k \geq 0, \; k = 1, 2, \ldots, n$$
$$s_d^- \geq 0, \; b = 1, 2, \ldots, p$$
$$s_g^+ \geq 0, \; g = 1, 2, \ldots, o$$

The first multiplicative form to the solve problem is as follows:

$$min\xi - \varepsilon \left(\sum_{d=1}^{m} s_d^- + \sum_{g=1}^{s} s_g^+ \right)$$

S.t :

$$\sum_{e=1}^{n} x_{de}\alpha_e + s_d^- = \xi x_{i0}, \; d = 1, 2, \ldots, p$$
$$\sum_{e=1}^{n} y_{ge}\alpha_e - s_g^+ = y_{g0}, \; g = 1, 2, \ldots, o \tag{21}$$
$$\sum_{k=1}^{n} \alpha_k = 1 \alpha_k \geq 0, \; k = 1, 2, \ldots, n$$
$$s_d^- \geq 0, \; b = 1, 2, \ldots, p$$
$$s_g^+ \geq 0, \; r = 1, 2, \ldots, o$$

A second multiplier form of the linear program is expressed as:

$$\max_{c.a,a_0} \xi_B = a^Z y_0 - a_0$$

S.t :

$$c^Z x_0 = 1$$
$$a^Z y_e - c^Z x_e - a_0 \leq 0, \; e = 1, 2, \ldots, n \tag{22}$$
$$c \geq 0$$
$$a \geq 0$$

As mentioned in Formula (14), in this case Z and u are vectors, and the scalar v_0 may be positive or negative or zero. The dual program for the equivalent BCC fractional program (12) can be obtained as follows:

$$\max_{c.a} \xi = \frac{a^Z y_0 - a_0}{c^Z x_0}$$

S.t :

$$\frac{a^Z y_e - a_0}{c^Z x_e} \leq 1, \; e = 1, 2, \ldots, n \tag{23}$$
$$c \geq 0$$
$$a \geq 0$$

A BCC-efficient solution is a solution where if an optimal solution, DMU$_0$, $(\xi_B^*, s^{-*}, s^{+*})$ as solved in this two phase processes for model satisfies $\xi_B^* = 1$ and has no slack values where $s^{-*} = s^{+*} = 0$. Otherwise, the model would be considered as BCC-inefficient.

The improved activity, $(\xi^* x - s^{-*}, y + s^{+*})$, also can be claimed as BCC efficient. A DMU, which is a minimized input value for any input item, or a maximized output value for any output item, is BCC-efficient.

The output-oriented BCC model (BCC-O) is defined as the following:

$$\max \eta$$

S.t :

$$\sum_{e=1}^{n} x_{de}\alpha_e + s_d^- = \xi x_{d0}, \; b = 1,\, 2,\ldots,\, p$$
$$\sum_{e=1}^{n} y_{ge}\alpha_e - s_g^+ = \eta y_{g0}, \; g = 1,\, 2,\ldots,\, o \tag{24}$$
$$\sum_{k=1}^{n} \alpha_k = 1$$
$$\alpha_k \geq 0, \; k = 1,\, 2,\ldots,\, f$$

A multiplier form of the linear program (24) can be expressed as [25]:

$$\min_{c.a,c_0} a^Z y_0 - a_0$$

S.t :

$$a^Z y_0 = 1$$
$$c^Z x_e - a^Z y_e - a_0 \leq 0, \; e = 1,\, 2,\ldots,\, n \tag{25}$$
$$c \geq 0$$
$$a \geq 0$$

In the envelopment model, the Z_0 is the scalar combined with $\sum_{k=1}^{n} \alpha_k = 1$. Conclusively, the equivalent fractional programming formulation for the BCC model was achieved by the authors (25):

$$\min_{c.a,c_0} \frac{c^Z x_0 - c_0}{a^Z y_0}$$

S.t :

$$\frac{c^Z x_e - a_0}{a^Z y_e} \leq 1, \; e = 1,\, 2,\ldots,\, n \tag{26}$$
$$c \geq 0$$
$$a \geq 0$$

3.2.3. Slacks Based Measure Model (SBM Model)

The SBM model was developed by Tone, which has three elements, input-oriented, output-oriented. Input-Oriented SBM (SBM-I-C).

The following model can be defined as the Input-oriented SBM under constant-returns-to-scale-assumption:

$$\rho_I^* = \min_{\alpha,\, s^-,s^+} 1 - \frac{1}{m} \sum_{d=1}^{m} \frac{s_d^-}{x_{dh}}$$

S.t :

$$x_{dc} = \sum_{e=1}^{m} x_{dc}\alpha_d + s_d^-, \; d = 1,\, 2,\ldots,\, p$$
$$y_{gc} = \sum_{e=1}^{m} y_{gc}\alpha_e - s_g^+, \; g = 1,\, 2,\ldots,\, o \tag{27}$$
$$\alpha_e \geq 0, \; k \, (\forall j), \; s_d^- \geq 0 \, (\forall e), s_g^+ \geq 0 \, (\forall e)$$

4. Case Study

Solar power in Vietnam belongs to the emerging energy industry group, which is engaged in the development of the world's renewable energy sources, the import of science and technology, and meets the demand for developing power sources. When large hydropower sources are fully exploited, small hydroelectricity will not guarantee benefits compared to the environmental damage. Vietnam, on the other hand, has great potential for solar and wind power, due to its proximity to the equator and the existence of dry, sunny regions like the southern central provinces. The study results are shown in Figure 6. Therefore, solar power, together with wind power, is being encouraged to develop by the State of Vietnam, reflected in the Prime Minister's Decision No. 2068/QD-TTg of 25 November 2015, approving the Renewable Energy Development Strategy. Vietnam's electricity generation until 2030, with a vision of 2050, ensures the development of electricity sources when stopping nuclear power projects and reducing fossil-fired thermal power plants.

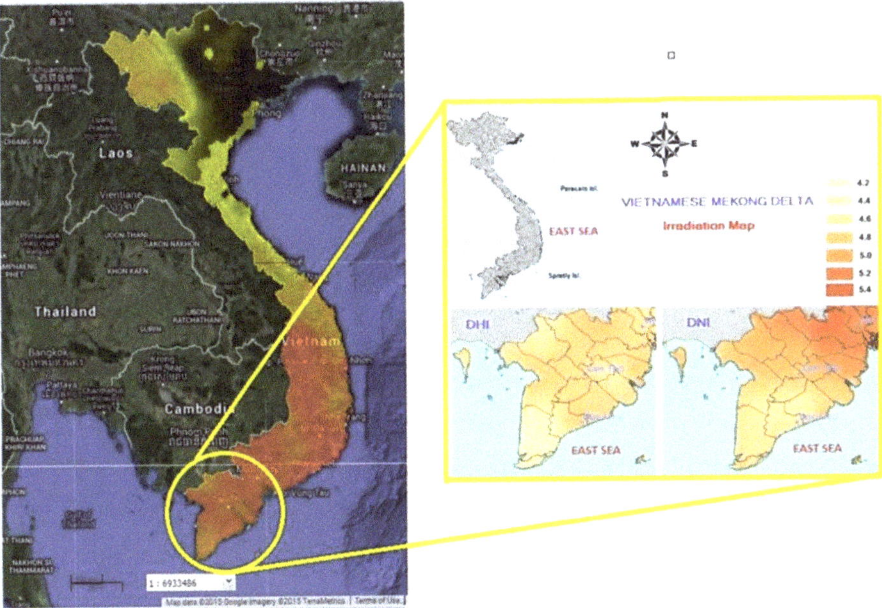

Figure 6. Irradiation Map of Mekong Delta, Vietnam [27].

With the growth of economic sectors in general and agriculture, the demand for electricity in the Mekong Delta provinces is increasing. However, fossil fuel for electricity generation is declining, so the development of renewable energy is one of the sustainable solutions for the Mekong Delta to have electricity to ensure domestic living and economic development.

In this research, the authors present an integrated approach for the site of solar plants in Mekong Delta, Vietnam, by using data envelopment analysis (DEA) and Fuzzy Analytical Network Process (FANP). In the first stage, DEA model is applied to select some potential location, then FANP model is used for ranking these potential locations. The authors collected data from thirteen locations in Mekong Delta, Vietnam, which can invest in solar power plants, as shown in Table 3.

Table 3. List of the 13 locations identified in Mekong Delta, Viet Nam.

No	Locations	Symbols	No	Locations	Symbols
1	An Giang	DEL01	8	Kien Giang	DEL08
2	Bac Lieu	DEL02	9	Long An	DEL09
3	Ben Tre	DEL03	10	Soc Trang	DEL10
4	Ca Mau	DEL04	11	Tien Giang	DEL11
5	Can Tho	DEL05	12	Tra Vinh	DEL12
6	Dong Thap	DEL06	13	Vinh Long	DEL13
7	Hau Giang	DEL07	-	-	-

DEA model is a quantitative technique that determines the relative effectiveness of multiple inputs and outputs decision makers. Halasah et al. [28] employed life-cycle assessment to evaluate the energy-related impacts of PV systems at different scales of integration. The input parameters included panel efficiency, temperature coefficient, shading losses, ground cover ratio and latitude, and the input data included hourly solar radiation, wind speed and temperature. Wang and Amy [9,14] using DEA model for ranking potential location for building solar power plant. In their research, the output data included sunshine hour and elevation. Due to the information accessibility of various sites and the importance of various factors, we select two inputs and two outputs for the quantitative factors. The two inputs are temperature (I1) and wind speed (I2). The two outputs are sunshine hours (O1) and elevation (O2) [14,20]. The definition of the inputs and outputs are defined in Table 4 [20]. Raw data of inputs and outputs of DMUs are demonstrated in Figure 7.

Table 4. Definition of inputs and outputs.

Factors	Definition
	Inputs
Temperature (I_1)	This is a measure of how high or low the heat radiation of the current environment is. Radziemska [29] discussed that should the temperature increase, the overall output power and the conversion efficiency of the PV module would decrease. A solar panel that receives more sunlight would lead to an increase in temperature on the panel, hence a decrease in the conversion efficiency. This decrease can vary from 10% to 25% depending on the location and material of the equipment. [30]
Wind speed (I_2)	Wind is the current of gas particles that flows randomly. Depending on the strength of the current, small particles from solid sediments to suspended objects could obstruct and affect the solar panels and other equipment. The wind can cause erosion and potential failures in operation for solar power plants.
	Outputs
Sunshine hours (O_1)	An indicator that measures the average duration of sunshine in a time period (here, annually) for a given location. This period is when direct solar irradiance shows a measurement higher than 120 W/m^2 [31]. A higher output power would be produced should a higher sunshine hour be recorded.
Elevation (O_2)	The geographical height above sea level of a given location (here, meters). If the geographical elevation is high, the distance between the ground to the sun is decreased, which means the solar radiant takes a shorter time to reach the ground. Additionally, lower distance means a lower dispersion of solar radiation, so the intensity is projected to be higher. Higher intensity yields higher solar energy output. Panjwani and Narejo discussed how elevation generated a 7–12% increase in power by testing 3 solar panels at a 27.432 m elevation [32].

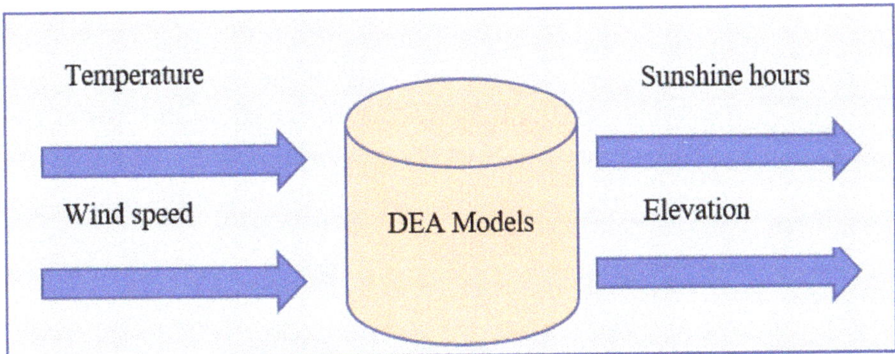

Figure 7. Inputs and outputs of DEA models.

For applied DEA model, some additional data about 13 locations in Mekong Delta, Vietnam are show in Table 5.

Table 5. Raw data of DEA model.

DMUs	Temperature (I_1)	Wind Speed (I_2)	Sunshine hours (O_1)	Elevation (O_2)
DEL01	27.20	1.70	2589	2.00
DEL02	27.30	7.20	2700	1.50
DEL03	27.00	1.00	2341	2.00
DEL04	26.80	1.30	2300	1.00
DEL05	26.90	2.80	2672	2.00
DEL06	27.40	2.50	2500	2.00
DEL07	27.30	7.00	2700	0.80
DEL08	27.50	6.00	2563	1.20
DEL09	27.70	2.80	2800	1.60
DEL10	26.80	6.40	2292	0.90
DEL11	27.00	1.70	2645	1.00
DEL12	27.60	6.80	2556	0.80
DEL13	28.00	1.40	2400	1.60

To select some potential sites in Mekong Delta, Vietnam, there are several DEA models, including SBM-I-C; CCR-I; BCC-O; CCR-O and BCC-I, applied in this step. The results of the DEA models are shown in Table 6.

Table 6. Ranking results from some DEA models.

Locations	DEA Approaches				
	SBM-I-C	CCR-I	BCC-O	CCR-O	BCC-I
DEL01	1	1	1	1	1
DEL02	0.6546	0.9784	0.9868	0.9784	0.9918
DEL03	1	1	1	1	1
DEL04	0.8728	0.936	0.9999	0.936	1
DEL05	1	1	1	1	1
DEL06	0.7767	0.9824	1	0.9824	0.9824
DEL07	0.6451	0.9784	0.9868	0.9784	0.9918
DEL08	0.6159	0.922	0.9259	0.9220	0.9771
DEL09	1	1	1	1	1
DEL10	0.5521	0.8461	0.9961	0.8461	1
DEL11	1	1	1	1	1
DEL12	0.5935	0.9162	0.9181	0.9162	0.9735
DEL13	0.8604	0.9325	0.9595	0.9325	0.962

As the results in Table 6 show that there are seven DMUs that are potential locations for building a solar power plant in Mekong Delta, Vietnam including DEL01, DEL03, DEL05, DEL09 and DEL11. These DMUs will be evaluated in the next step of this research by using the FANP model.

In the final stage, all the potential locations will be ranked by the FANP model. Some criteria affecting the location selection are shown in Figure 8.

Figure 8. Main criteria and sub-criteria in Fuzzy Analytical Network Process (FANP) model.

Fuzzy comparison matrix of EC from the FANP model is shown in Table 7.

Table 7. Fuzzy comparison matrices for Economic factor (EC).

	Technological (TL)	Site Characteristics (ST)	Environmental (EL)	Social (SL)
TL	(1,1,1)	(5,6,7)	(1,2,3)	(3,4,5)
ST	(1/7,1/6,1/5)	(1,1,1)	(1/3,1/4,1/5)	(2,3,4)
EL	(1/3,1/2,1)	(5,4,3)	(1,1,1)	(4,5,6)
SL	(1/5,1/4,1/3)	(1/4,1/3,1/2)	(1/6,1/5,1/4)	(1,1,1)

The fuzzy numbers were converted to real numbers by using the TFN. During the defuzzification, the authors obtain the coefficients $\alpha = 0.5$ and $\beta = 0.5$. In this, α represents the uncertain environment conditions, and β represents whether the attitude of the evaluator is fair.

$$g_{0.5,0.5}(\overline{a_{EL,SL}}) = [(0.5 \times 4.5) + (1 - 5.5) \times 2.5] = 5$$

$$f_{0.5}(L_{EL,SL}) = (5 - 4) \times 0.5 + 4 = 4.5$$

$$f_{0.5}(U_{SL,SL}) = 6 - (6 - 5) \times 0.5 = 5.5$$

$$g_{0.5,0.5}(\overline{a_{SL,EL}}) = 1/5$$

The remaining calculations for other criteria are similar to the above calculation. The real number priority when comparing the main criteria pairs is shown in Table 8.

Table 8. Real number priority.

	TL	ST	EL	SL
TL	1	6	2	4
ST	1/6	1	1/4	3
EL	1/2	4	1	5
SL	1/4	1/3	1/5	1

The following are used for calculating the maximum individual value:

$$P1 = (1 \times 6 \times 2 \times 4)^{1/4} = 2.63$$

$$P2 = (1/5 \times 1 \times 1/4 \times 3)^{1/4} = 0.62$$

$$P3 = (1/2 \times 4 \times 1 \times 5)^{1/4} = 1.78$$

$$P4 = (1/4 \times 1/3 \times 1/5 \times 1)^{1/4} = 0.36$$

$$\sum Q = 5.39$$

$$\omega_1 = \frac{2.63}{5.39} = 0.49$$

$$\omega_2 = \frac{0.62}{5.39} = 0.12$$

$$\omega_3 = \frac{1.78}{5.39} = 0.33$$

$$\omega_4 = \frac{0.36}{5.39} = 0.07$$

$$\begin{bmatrix} 1 & 6 & 2 & 4 \\ 1/6 & 1 & 1/4 & 3 \\ 1/2 & 4 & 1 & 5 \\ 1/4 & 1/3 & 1/5 & 1 \end{bmatrix} \times \begin{bmatrix} 0.49 \\ 0.12 \\ 0.33 \\ 0.07 \end{bmatrix} = \begin{bmatrix} 2.15 \\ 0.49 \\ 1.41 \\ 0.30 \end{bmatrix}$$

$$\begin{bmatrix} 2.15 \\ 0.49 \\ 1.41 \\ 0.30 \end{bmatrix} / \begin{bmatrix} 0.49 \\ 0.12 \\ 0.33 \\ 0.07 \end{bmatrix} = \begin{bmatrix} 4.39 \\ 4.08 \\ 4.27 \\ 4.29 \end{bmatrix}$$

Based on the number of main criteria, the authors get n = 5; λ_{max} and CI are calculated as follows:

$$\lambda_{max} = \frac{4.39 + 4.08 + 4.27 + 4.29}{4} = 4.23$$

$$CI = \frac{\lambda_{max} - n}{n - 1} = \frac{4.23 - 4}{4 - 1} = 0.077$$

To calculate CR value, we get RI = 0.9 with n = 4.

$$CR = \frac{CI}{RI} = \frac{0.077}{0.9} = 0.08556$$

As CR = 0.08556 $\leq$ 0.1, we do not need to re-evaluate.

5. Discussion

There have been many studies presenting MCDM models using fuzzy set theory to evaluate and select the location of a renewable energy plant, particularly solar energy. However, the literature review shows that most of these studies only use economic, or technological, or site characteristics, or environmental or social standards to assess potential locations. There are few studies that simultaneously use economic, technological, site characteristics, environmental and social factor standards in the process of location selection. In addition, there seems to be no research on developing an MCDM model to choose a location for a solar power plant in the Mekong Delta.

This is a reason the author proposed an MCDM model including data envelopment analysis (DEA) and fuzzy analytical network process (FANP) for solar power plant location selection in Mekong Delta, Vietnam in this work. In the first step, some potential location is selected from 13 location in Mekong Delta by using the DEA model. Then, the FANP model is applied for rank potential sites. In a renewable energy project, location decision is the most importance activity; location decision is always a multicriteria decision making factor, as the decision maker must consider both quantitative and qualitative factors. In any renewable energy project, location decision is a very important decision, decision maker must consider qualitative and quantitative factors. Thus, solar power plant location selection is a multicriteria decision making (MCDM). For selection optimal location, the proposed model considered five main factors: economic factors (construction cost, operation and management cost, new feeder cost), technological factors (distance from major road, distance from power network, potential demand), site characteristics factors (ecology, elevation, approachability), environmental factors (temperature, sunshine hours, humidity) and social factors (supports mechanisms, protection law and legal regulatory compliance) As per the results shown in Table 9, Ben Tre (DEL03) is the optimal solution for investing solar power plants in Mekong Delta, Vietnam.

Table 9. Final ranking from the FANP model.

No	Location-Symbols	Score	Ranking Order
1	An Giang-DEL01	0.226	3
2	Ben Tre-DEL03	0.246	1
3	Can Tho-DEL05	0.238	2
4	Long An-DEL09	0.107	4
5	Tien Giang-DEL11	0.183	5

6. Conclusions

Facing energy demand challenges to ensure sustainable economic growth, especially the exhaustion of fossil fuels, environmental pollution, and climate change. Incentive policies to take advantage of renewable energy supply opportunities in Vietnam. This approach to taking advantage of this new energy source not only contributes to timely supply of the energy needs of the society but also helps to save electricity and reduce environmental pollution.

However, when considering investing in solar power plants, decision makers must consider many qualitative and quantitative factors including economic, technological, site characteristics, environmental and social factors. Therefore, this problem is tackled as a Multi-Criteria Decision Making (MCDM) problem that requires that both quantitative and qualitative information are considered. This paper presented an integrated approach for site of solar plants by data envelopment analysis (DEA) and Fuzzy Analytical Network Process (FANP). Furthermore, this integrated approach, incorporating the most relevant parameters of solar plants, is introduced.

The contributions of this research include a fuzzy multicriteria decision-making (F-MCDM) approach for solar power plant site selection in Vietnam. This research also utilizes the evolution of a new approach that is flexible and practical for the decision-maker and provides useful guidelines for solar power plant location selection in many countries around the world.

For further studies regarding this topic, the study can be expanded to other MCDM approaches such as VIKOR, ELECTRE I, ELECTRE III.

Author Contributions: Conceptualization, C.-N.W., T.P.H., V.T.N. and S.T.H.; Data curation, V.T.H.V., T.P.H. and V.T.N.; Formal analysis, C.-N.W., V.T.H.V., T.P.H., V.T.N. and S.T.H.; Funding acquisition, C.-N.W. and V.T.H.V.; Investigation, C.-N.W., T.P.H., V.T.N. and S.T.H.; Methodology, C.-N.W., V.T.H.V., T.P.H. and V.T.N.; Project administration, V.T.H.V., T.P.H. and V.T.N.; Resources, V.T.N.; Visualization, T.P.H. and S.T.H.; Writing—original draft, V.T.H.V., T.P.H.,V.T.N. and S.T.H.; Writing—review and editing, C.-N.W., V.T.H.V. and V.T.N. All authors have read and agreed to the published version of the manuscript.

Funding: This research received no external funding.

Conflicts of Interest: The authors declare no conflict of interest.

References

1. Renewable Capacity Growth between 2019 and 2024 by Technology. Available online: https://www.iea.org/data-and-statistics/charts/renewable-capacity-growth-between-2019-and-2024-by-technology (accessed on 22 March 2020).

2. Ưu Khuyết Điểm Của Năng Lượng Mặt Trời. Available online: https://www.iea.org/data-and-statistics/charts/renewable-capacity-growth-between-2019-and-2024-by-technology (accessed on 10 March 2020).

3. World Summit Outcome, Resolution A/60/1, adopted by the General Assembly on 15 September 2005. Available online: https://www.un.org/en/development/desa/population/migration/generalassembly/docs/globalcompact/A_RES_60_1.pdf (accessed on 5 March 2020).

4. MRC Strategic Environmental Assessment (SEA) of Hydropower on the Mê Công Mainstream. Available online: http://www.mrcmekong.org/assets/Publications/Consultations/SEA-Hydropower/SEA-FR-summary-13oct.pdf (accessed on 5 March 2020).

5. Đồng Bằng Sông Cửu Long Trước Nguy cơ lớn Từ Nhiệt điện Than. Available online: https://www.thiennhien.net/2018/06/22/dong-bang-song-cuu-long-truoc-nguy-co-lon-tu-nhiet-dien-than (accessed on 22 March 2020).

6. Gorjian, S.; Hashjin, T.T.; Ghobadian, B. Estimation of mean monthly and hourly global solar radiation on surfaces tracking the sun: Case study: Tehran. In Proceedings of the Second Iranian Conference on Renewable Energy and Distributed Generation, Tehran, Iran, 6–8 March 2012; pp. 172–177.

7. Điện Mặt Trời Phát Triển Mạnh Ở Đồng Bằng Sông Cửu Long. Available online: http://ceid.gov.vn/dien-mat-troi-phat-trien-manh-o-dong-bang-song-cuu-long (accessed on 22 March 2020).

8. Gloal Solar Atlas. Available online: https://globalsolaratlas.info/map?c=-29.22889,-1.40625,2 (accessed on 22 March 2020).

9. Wang, C.N.; Nguyen, V.T.; Thai, H.T.N.; Duong, D.H. Multi-Criteria Decision Making (MCDM) Approaches for Solar Power Plant Location Selection in Viet Nam. *Energies* **2018**, *11*, 1504. [CrossRef]

10. Kengpol, A.; Rontlaong, P.; Tuominen, M. A decision support system for selection of solar power plant locations by applying fuzzy ahp and topsis: An empirical study. *J. Softw. Eng. Appl.* **2013**, *6*, 470–481. [CrossRef]

11. Azadeh, A.; Ghaderi, S.F.; Maghsoudi, A. Location optimization of solar plants by an integrated hierarchical DEA PCA approach. *Energy Policy* **2008**, *36*, 3993–4004. [CrossRef]

12. Thongpun, A.; Nasomwart, S.; Peesiri, P.; Nananukul, N. Decision support model for solar plant site selection. In Proceedings of the 2017 IEEE International Conference on Smart Grid and Smart Cities (ICSGSC), Singapore, 23–26 July 2017; pp. 50–54.

13. Azadeh, A.; Sheikhalishahi, M.; Asadzadeh, S.M. A flexible neural network-fuzzy data envelopment analysis approach for location optimization of solar plants with uncertainty and complexity. *Renew. Energy* **2011**, *36*, 3394–3401. [CrossRef]

14. Lee, A.H.I.; Kang, H.-Y.; Lin, C.-Y.; Shen, K.-C. An integrated decision-making model for the location of a pv solar plant. *Sustainability* **2015**, *7*, 13522–13541. [CrossRef]

15. Sozen, A.; Mirzapour, A.; Çakir, M.T. Selection of the best location for solar plants in Turkey. *J. Energy S. Afr.* **2015**, *26*, 52–63. [CrossRef]

16. Azadeh, A.; Rahimi-Golkhandan, A.; Moghaddam, M. Location optimization of wind power generation–transmission systems under uncertainty using hierarchical fuzzy DEA: A case study. *Renew. Sustain. Energy Rev.* **2014**, *30*, 877–885. [CrossRef]
17. Yokota, S.; Kumano, T. Mega solar optimal allocation using data envelopment analysis. *Electr. Eng. Jpn.* **2013**, *183*, 24–32. [CrossRef]
18. Chabuk, A.; Al-Ansari, N.; Hussain, H.M. Landfill sites selection using MCDM and comparing method of change detection for Babylon Governorate. *Environ. Sci. Pollut. Res.* **2019**, *26*, 35325–35339. [CrossRef] [PubMed]
19. Chakraborty, R.; Ray, A.; Dan, P. Multi criteria decision making methods for location selection of distribution centers. *Int. J. Ind. Eng. Comput.* **2013**, *4*, 491–504. [CrossRef]
20. Lee, A.H.; Kang, H.Y.; Liou, Y.J. A hybrid multiple-criteria decision-making approach for photovoltaic solar plant location selection. *Sustainability* **2017**, *9*, 184. [CrossRef]
21. Azadeh, A.; Ghaderi, S.F.; Nasrollahi, M.R. Location optimization of wind plants in Iran by an integrated hierarchical data envelopment analysis. *Renew. Energy* **2011**, *36*, 1621–1631. [CrossRef]
22. Fang, L.; Li, H. Multi-criteria decision analysis for efficient location-allocation problem combining DEA and goal programming. *RAIRO Oper. Res.* **2015**, *49*, 753–772. [CrossRef]
23. Dehghani, E.; Jabalameli, M.S.; Pishvaee, M.S.; Jabarzadeh, A. Integrating information of the efficient and anti-efficient frontiers in DEA analysis to assess location of solar plants: A case study in Iran. *J. Ind. Syst. Eng.* **2018**, *11*, 163–179.
24. Mostafaeipour, A.; Sedaghat, A.; Qolipour, M.; Rezaei, M.; Arabnia, H.R.; Saidi-Mehrabad, M.; Alavi, O. Localization of solar-hydrogen power plants in the province of Kerman. *Adv. Energy Res.* **2017**, *5*, 179.
25. Azadeh, A.; Ghaderi, S.F.; Maghsoudi, A. Location optimization of solar plants by an integrated multivariable DEA-PCA method. In Proceedings of the 2006 IEEE International Conference on Industrial Technology, Mumbai, India, 15–17 December 2006; pp. 1867–1872.
26. Lee, S.K.; Mogi, G.; Hui, K.S. A fuzzy analytic hierarchy process (AHP)/data envelopment analysis (DEA) hybrid model for efficiently allocating energy R&D resources: In the case of energy technologies against high oil prices. *Renew. Sustain. Energy Rev.* **2013**, *21*, 347–355.
27. Tuan, L.A. An overview of the renewable energy potentials in the Mekong river Delta, Vietnam. *Can Tho Univ. J. Sci.* **2016**, *4*, 70–79.
28. Halasah, S.A.; Pearlmutter, D.; Feuermann, D. Field installation versus local integration of photovoltaic systems and their effect on energy evaluation metrics. *Energ. Pol.* **2013**, *52*, 462–471. [CrossRef]
29. Radziemska, E. The effect of temperature on the power drop in crystalline silicon solar cells. *Renew. Energy* **2003**, *28*, 1–12. [CrossRef]
30. Solar Panel Temperature Affects Output—Here's What You Need to Know. Available online: http://www.solar-facts-and-advice.com/solar-panel-temperature.html (accessed on 5 March 2020).
31. Guide to Meteorological Instruments and Methods of Observation. Available online: http://www.wmo.int/pages/prog/gcos/documents/gruanmanuals/CIMO/CIMO_Guide-7th_Edition2008.pdf (accessed on 5 March 2020).
32. Panjwani, M.; Narejo, D. Effect of altitude on the efficiency of solar panel. *Int. J. Eng. Res. Gen. Sci.* **2014**, *2*, 461–463.

MDPI

St. Alban-Anlage 66

4052 Basel

Switzerland

Tel. +41 61 683 77 34

Fax +41 61 302 89 18

www.mdpi.com

Energies Editorial Office

E-mail: energies@mdpi.com

www.mdpi.com/journal/energies